Climate Change Reality Check

Global Warming and Changes in Earth's Weather Patterns, Facts about Billion-Dollar Disasters and Solutions Using Artificial Intelligence

Freeman Publishing

eBook ISBN 978-1-963333-04-6

Paperback ISBN 978-1-963333-05-3

Hardback ISBN 978-1-963333-06-0

Table of Contents

Introduction

What do you do when there is nowhere to go? Two years after Hurricane Harvey hit, on August 25, 2017, the survivors recalled looking down the streets and seeing gaps where their neighbors' houses used to be, like broken teeth. Some of those gaps are still there. In the storm's aftermath, streets were choked with debris—cars, trailers, household goods, electricity poles, knots of utility cable, tarmac slabs, and downed trees. Some achingly recall watching the furniture they had accumulated over decades go onto the curb because it was unusable. However, it is the personal losses that make people tear up more than the ruined homes that smell of putrid water—irreplaceable things like the photos of your kids growing up, the chair your grandfather made for you, and the family Bible. These days, flood insurance is slow to pay and it can take weeks to receive payouts. By then, the builders are overwhelmed, so your house rots quietly in the coastal sun while you find temporary digs or move in with relatives. Some inhabitants decide they might be better off moving away from their former idyllic place in the sun.

Then there's post-traumatic stress syndrome (PTSD). Every time it starts to rain, especially when the rain is torrential, the survivors of hurricanes, tropical storms, and unexpected floods start panicking. They can't breathe, they can't sleep, they pace their new homes, and watch for water creeping beneath the doors. Children ask fearfully, "Is it going to stop, Mommy?" Rain becomes a four-letter word, a traumatic experience that lives on in the subconscious, allied to the horror of waking up in the early hours with muddy water knee-deep in your house, worrying about children and pets. The terror of being at the mercy of wind and water never quite leaves you even after you've rebuilt your life.

There's good news too: the law enforcement officials who gave people hope and news after Hurricane Harvey when there was no cell service for days, people who cooked for hordes of hungry and cold survivors, while other storm victims scoured their sodden houses for clothing and blankets to share with others. Some threw themselves into community work, locating missing pets and people, helping with insurance claims, and channeling donations to people who had lost everything they owned. The cost is incalculable, especially if you're a survivor.

The Cost of the Climate Crisis

According to the North Carolina Institute for Climate Studies (Maycock, 2022), climate change and related weather events are costing the US more each year. Between 1980 and 2021, the country experienced over 300 weather-related disasters where overall damages for each one exceeded $1 billion. The total cost for all these disasters is estimated at over $2 trillion, with every US state, Puerto Rico, and the Virgin Islands all having experienced at least one $1 billion-dollar disaster since 1980, with the Southeastern US particularly hard hit.

As the Earth's climate has warmed, more people are experiencing episodes of extreme weather, including record-breaking hurricanes, floods, droughts, heat waves, and hail storms. Severe weather is determined by historic weather records and needs to be more extreme than similar weather events previously experienced in a specific area. In 2020, there were 22 such events, while 2021 saw 20 occurrences. In 2021, the damage from Hurricane Ida alone, which hit Cuba, Jamaica, and the Cayman Islands on August 25 that year, amounted to $75 billion. By October of 2023, the US had already experienced 24 severe weather events, with cumulative damages totaling $57.6 billion (Oladipo, 2023).

Climate change increases the chances of severe weather in several ways. It intensifies heat waves, which raise evaporation levels, thereby exacerbating droughts. Drought dries out vegetation, leading to a higher incidence of wildfires and increasing their intensity. A hotter planet releases more water vapor into the atmosphere, so torrential rain and severe snowstorms become more common. The oceans are also warming, and so is the atmosphere above them, driving more severe storms. Rising sea levels due to melting polar ice caps are causing storm surges that inundate coastlines and decimate infrastructure.

These natural disasters are increasingly affecting people. In the US in 2021, nearly 1,000 people died as a result of extreme weather events. In addition, families spent more on necessities as Hurricanes Harvey and Irma approached. Following these disasters, the economies in the affected areas contracted by as much as 20%. It sometimes takes families years to make up for these losses. Having to rely on the Federal Emergency Management Agency (FEMA) for relief has a negative effect on credit scores, especially for lower-income households. Federal government funding to provide relief for victims in the wake of hurricane-related disasters now accounts for around 60% of damages costs. Research indicates that the higher the wind speed, the more the economic contraction (The White House, 2022).

As climate change increases the duration, intensity, and frequency of extreme weather events, this is making it more difficult for regional economies to recover from weather-related disasters. Newer studies are indicating that economic growth flattens as temperatures rise. Not only that, climate-related economic damage tends to be cumulative. The new models indicate that this is likely to be ten times more severe than previously thought (The White House, 2022).

Climate Change Impacts

The statistics tell an alarming story—one we're all familiar with. In the last century, global temperatures have risen by around 35.24 °F (1.8 °C) on average. Since the late 1970s, Arctic sea ice coverage has plunged by 40%, while the density of most well-known glaciers has reduced by over 60 feet. Atmospheric carbon dioxide has increased by 25% since the late 1950s (NOAA, 2021).

Climate change is not a linear problem. One change in a complex system can produce unexpected ripple effects. For example, a drought might negatively impact crop production and accordingly human health due to food shortages or famine. Flooding can cause waterborne diseases to proliferate while damaging infrastructure and even natural systems.

The impacts of global warming can affect multiple systems. Heavier rainfall events are leading to flooding across many parts of the US. In contrast, droughts, especially in the Southwest, have become more common. Because it's hotter, more water is needed to irrigate crops, which dry out faster, and people also consume more to cool off. This is putting an increasing strain on water supplies. Snowpack is an important freshwater source for many, but as climate change reduces the amount of winter snowfall, less water is available for communities that rely on this resource.

Producing food is becoming more difficult. Farmers need to contend with higher temperatures, weather extremes, droughts, floods, water stress, and new diseases and pests that have migrated from elsewhere due to climate change. Agricultural workers may suffer from heat stroke and dehydration. Livestock could succumb to extreme heat or inclement weather.

Heat waves, hurricanes, droughts, flooding, and diseases borne by insects can affect human health. Children, the elderly, people from lower-income groups, and communities of color are particularly likely to experience adverse health effects as a result of climate change.

Climate change is disrupting ecosystems and living organisms everywhere. More adaptable, generalist species are surviving. Flowers might bloom earlier, while animals and birds can migrate to new regions. However, the changes are so rapid that plants and animals are struggling to adapt. Living creatures also depend on one another for survival in a timeless circle of life. If a migratory bird species arrives at its summer breeding grounds but an anomalous weather event leads to a heavy late snowfall, it won't be able to find food, let alone breed. Some birds are arriving after the caterpillars they used to depend on to feed their chicks have long turned into butterflies, and flowers are blooming out of sync with their pollinators. All this disrupts living systems, putting them at risk.

The ocean has been absorbing the excess carbon we are pumping into the atmosphere for decades and is now acidifying. This acidification is affecting marine life. Some organisms are no longer able to form protective shells, for example. Due to a combination of increased heat, melting glaciers, and disappearing sea ice, sea levels are rising inexorably. Across the world, coral reefs are bleaching as waters warm. Hurricanes are destroying reefs while rising seas and storm surges smother them in sediment. Coral reefs are incredibly diverse ecosystems but they're fast disappearing in a warming world.

Infrastructure like roads, buildings, bridges, ports, and much else is being impacted by climate change. Unprecedented snowfalls make roads impassable, floods wash away tarmac and create sinkholes, ice breaks power lines, and we consume more energy for both heating and cooling.

Coastal living has increased in popularity over the years and 40% of the US population now lives on the coast. These areas are a locus for climate change impacts, including sea level rise, storm surges, and seawater entering freshwater aquifers. Some communities might be underwater by 2100 and there are already discussions concerning managing the retreat from the ocean (NOAA, 2021). Many communities are not yet prepared for a climate-disrupted world.

Climate Change: A Serious Concern for Young People

A survey conducted in early 2021 found that 83% of Gen Z Americans (people aged 14–24) were concerned about environmental issues, especially climate change, due to its potentially negative effects on their lives. A third of respondents had had their exercise routines disrupted by natural disasters at least once in the previous five years. Many were found to be suffering from eco-anxiety, a condition similar to PTSD. Most believed that environmental degradation had affected their health, and were particularly concerned about air and water pollution.

But all is not lost. Many celebrities are working to improve environmental awareness. Leonardo DiCaprio, Greta Thunberg, Mark Ruffalo, Prince Harry, and Jane Fonda, to name only a few, are constantly using their enormous fan bases to raise awareness of the environmental crisis and how climate change is threatening the planet. Greta Thunberg, a well-known Swedish climate change activist, has created a global movement among today's young people.

The impacts of these campaigns have been noticeable, especially among millennials and Gen Z, who have grown up witnessing the devastation wreaked by climate change—and in some cases experiencing it firsthand. They are incorporating imperfect, misshapen vegetables into their diets, eating less meat, and supporting brands and employers who advocate for the environment. About 75% of Gen Z are frightened of the future, according to a recent medical study. Even as they struggle to save the planet, many are experiencing climate anxiety. Because of the surging interest and demand for qualifications and employment in the environmental and sustainability sectors, more college campuses are offering courses in these disciplines (Sparks & Honey, n.d.).

Why Read This Book?

Imagine if everything could go back to the way it was, only better. The sky is a sapphire blue dome, filled with fluffy clouds. No smog obscures it—the air is clean and pure. Sparkling streams meander through lush, green forests teeming with life.

Walking along the paths, you find wildflower clusters, ferns, old wooden bridges, frogs, and deer. You can cup your hands and drink the water; there's no pollution here. As the day winds down, you take your canoe out on the calm waters of a natural lake that mirrors the sky. It's quiet, the only sound is the faint dip of your paddles, a fish jumping, or a bald eagle calling further downstream. The world is at rest—and so are you.

There is already such a place. Florida's Babcock Ranch is a hurricane-proof development that might be the model for the world we want. When Hurricane Ian left a trail of destruction in its wake, Babcock Ranch emerged relatively unscathed.

Built in 2018 to exceed local building codes, the significant town —five times the size of Manhattan—is a study in sustainability. Houses are surrounded by green spaces like forest trails, cycle paths, and golf courses. The attractive lakes where residents paddle kayaks are actually flood attenuation dams. The streets absorb excess water during heavy showers, while the community hall is reinforced to double as a shelter. An enormous solar farm powers the entire development—making Babcock Ranch the country's first solar-powered town—and generates enough energy for nearby communities as well.

During Hurricane Ian, the ferocious winds were strong enough to alarm residents but their fears were groundless; none of the houses lost power, internet, or access to clean water. In fact, the town offered emergency shelter to hard-hit locals who were nearby.

Babcock Ranch's resilience was the result of careful planning, even down to consulting old maps to determine how water moves across the site. Wetlands, which hold flood water and release it slowly, were retained. There are no structures in natural drainage easements so excess water can safely reach the Caloosahatchee River.

Coastal barrier island beaches and regional topography act as buffers during extreme weather events. Further protecting residents, power lines are buried underground. Babcock Ranch also has its own wastewater treatment plant, which means that residents have water both during and after severe storms.

You might be feeling anxious about the climate crisis. Perhaps you're a millennial or Generation Z, gazing down the barrel of a gun of a ruined world, exhausted by the fight to get individuals, communities, and corporations to understand. This isn't where the battle ends. There is hope. Today's technologies can enable us to have a better, more livable world, for ourselves and our children. In this book, you'll find out how climate change can be mitigated using cutting-edge technologies, like artificial intelligence, to create the world we want. It is my hope that this book will encourage and inspire you to move forward in making the world a better, more climate-proof place.

The Foundations of Climate Science

Did you know that our blue planet is warming at an unprecedented rate that hasn't been seen in the last 10,000 years? Since the first climate assessments in the 1970s, it has become increasingly apparent that human activities are the cause. As evidenced by ice core samples and similar data, Earth's climate has always been variable, with warmer periods alternating with ice ages. This is caused by our planet's elliptical orbit that alters our distance from the sun, changing the amount of solar energy reaching the surface.

The current warming trajectory falls outside these parameters, however. Since the Industrial Revolution in the mid-1800s, human activities have produced ever more of the gases that have the potential to heat the atmosphere excessively. This in turn affects conditions on the ground. Scientists have found evidence of this in ice core samples, tree rings, coral reefs, ocean sediments, and sedimentary rocks (Buis, 2023).

What is the Greenhouse Effect?

The media often portray the greenhouse effect as the disruptive force behind weather-related natural disasters. Nothing could be further from the truth. This natural phenomenon keeps us alive and ensures that the Earth can support life. Without it, surface temperatures on Earth would be similar to that of other planets in our solar system and outer space—around -0.4 °F (-18 °C). Held in place by gravity, the Earth's atmosphere traps the sun's heat, preventing solar radiation from escaping back into space, as it does on the moon, for example. Only a relatively small amount returns to space, often when sunshine encounters bright surfaces such as ice sheets and clouds. (Miller R.G. (2006) (Ye, 2023).

Certain gases present in the atmosphere absorb the sun's heat and disperse it in all directions, including toward the Earth's surface. Besides keeping the Earth warm enough to sustain life, this ensures that water remains liquid rather than being locked up as ice, further adding to our planet's biodiversity.

Greenhouse Gases and the Greenhouse Effect

The gases naturally present in the atmosphere that enable the greenhouse effect are water vapor, carbon dioxide, nitrous oxide, and methane—the so-called "greenhouse gases." In the past 250 years, human activities have raised atmospheric carbon dioxide levels so much that this has altered the balance of these gases in the atmosphere. Their heat-absorbing capacities have been exacerbated, and the excessive heat cannot dissipate into space. When there is more energy entering the atmosphere than leaving it, this creates an imbalance known as the enhanced greenhouse effect.

Carbon dioxide levels have risen by 31% since the late 1800s as a result of spiraling fossil fuel burning, deforestation, and cement production. Nitrous oxide emissions have grown by 17% due to commercial fertilizer usage, fuel and biomass consumption, and some industrial processes. Methane emissions are the real surprise—atmospheric levels have skyrocketed by 151%, driven by livestock agriculture, fossil fuel production, and proliferating landfill sites (Lemmons, 2023).

To make matters worse, we have created novel chemicals that act like greenhouse gases. These include the chlorofluorocarbons and hydrofluorocarbons used in refrigeration, nitrogen trifluoride used to manufacture semiconductors, and sulfur hexafluoride, a byproduct of electricity transmission.

There are also natural events that release carbon dioxide into the atmosphere, including:

- forest fires
- volcanic eruptions
- the decomposition of dead natural vegetation
- plant respiration (Ye, 2023)

Not all greenhouse gases have the same effect on the climate or Global Warming Potential (GWP), which is determined by:

1. their ability to absorb energy and disperse it; and
2. their atmospheric lifetime, which is how long the gas remains in the atmosphere before natural processes like chemical reactions remove it (Ye, 2023)

GWP is normally calculated over a century and is relative to carbon dioxide. Gases with high GWP values will usually warm the Earth more than the same amount of carbon dioxide. Others, such as methane, remain in the atmosphere for a longer time but may have a more intense warming effect.

Discussions around the role of greenhouse gas emissions in climate change center around carbon dioxide because this is not only more prevalent, it also accounts for much of the greenhouse effect. In 2021, carbon dioxide alone accounted for nearly 75% of the total warming effects of all anthropogenic greenhouse gases (Lindsey, 2023). At the time of the Industrial Revolution, the atmosphere contained around 280 parts per million (ppm) of carbon dioxide. By May 2023, the National Oceanic and Atmospheric Administration (NOAA) announced that atmospheric carbon dioxide was at a record high of around 424 ppm (Lemmons; Lindsey, 2023).

Climate research indicates that atmospheric carbon dioxide never prehistorically exceeded 300 ppm—levels last seen some three million years ago. Global surface temperatures were then 4.5–7.2 °F (6–7 °C) higher than in pre-industrial times, while sea levels were at least 16 feet higher than they were in 1900. If growing global energy demand continues to spur fossil fuel burning, then carbon dioxide levels in the atmosphere might even reach 800 ppm for the first time in 50 million years. We are indeed entering uncharted territory (Lindsey, 2023).

Implications for Earth's Atmosphere and Oceans

Atmospheric water vapor reacts to higher levels of carbon dioxide: the warmer the atmosphere becomes, the more water vapor it holds. This can triple the warming effects of greenhouse gases in the atmosphere. Water vapor therefore intensifies the effects of climate change (Buis, 2019).

This is particularly obvious in the Arctic where the atmosphere is becoming wetter and warmer. These conditions are increasing the rate at which sea ice is disappearing. This further amplifies warming as it exposes areas of dark ocean that absorb more heat as opposed to the white sea ice, which repels it.

The ocean is another major climate modifier and has absorbed around 25% of the excess carbon dioxide humans have emitted since the start of the industrial age, as well as 90% of the resulting heat (Shaw, 2023). Carbon dioxide dissolves in the ocean and reacts with seawater, making the sea more acidic. This process also reduces the amount of carbonate in the water, with serious implications for shellfish, corals, and calcareous plankton, which use this carbonate to form their shells and skeletons. Ocean acidification affects other organisms too—it reduces the ability of fish to identify predators, for instance. This has implications for marine food webs and coastal residents who depend on ocean resources for food and livelihoods.

There are indications that the carbon-storing capacity of the oceans is approaching its peak. Sea surface temperatures are rising to record levels, spawning tropical storms and hurricanes, and intensifying the El Nino Southern Oscillation (ENSO) that affects global rainfall patterns. The latter frequently causes deep droughts and torrential rains, leading to famine and floods.

As recently as August 2023, abnormally high sea surface temperatures turned a tropical storm into a hurricane. In the Hawaiian town of Lahaina in Maui, the winds fanned one of the worst urban firestorms in the US in 100 years (Shaw, 2023). Aging infrastructure exacerbated the disaster.

The build-up of carbon dioxide in the oceans has other, more serious implications. Natural increases in atmospheric carbon dioxide have warmed the Earth during various ice ages. However, when temperatures increased, the oceans outgassed their stored carbon dioxide as they heated up. This stored carbon then entered the atmosphere, amplifying the initial warming, so the ice age ended. This has implications for climate change, as the oceans may not store our excess carbon indefinitely.

Earth's Climate History

To understand the effects of human activity on the world's climate, it's necessary to consider the impact of historic climate patterns and natural phenomena.

Ice Ages and Interglacial Periods

Between 2.4 million and 11,500 years ago, the Earth's climate vacillated between freezing, glacial periods and warmer, interglacial times. During the ice ages, glaciers covered vast parts of the globe. These receded when the climate warmed again. The glacial periods lasted longer than the warmer ones and there were at least 17 cycles between the two. The last ice age ended 25,000 years ago and we are still in an interglacial age (Mammoth Discovery, 2019).

Natural Climate Drivers

This isn't the warmest the Earth has ever been. During the time of the dinosaurs, there was no sea ice, for example. And when ice ages are in progress it has been much colder. Several natural factors can potentially affect the climate.

The Sun

At the center of our solar system, the sun is the giant star that provides energy for life on Earth. It is a "variable star," so its brightness and the amount of energy it releases fluctuate in an 11-year cycle. The Earth warms and cools depending on how much energy the sun is producing (Seman, n.d.).

Earth's Orbit

The Earth's orbit changes over time, ranging from a complete circle to an ellipse. This affects the distribution of solar energy across the planet. The tilt of Earth's axis also changes, creating the seasons, while also determining average temperatures and ice ages. Variations in the Earth's orbit can take as long as 100,000 years to develop, while the cyclical tilts in the planet's axis take around 40,000 years to complete. The changes to our climate experienced over the last century cannot therefore be attributed to changes in Earth's orbit or the tilt of its axis (Seman, n.d.).

Volcanoes and Geologic Activity

Volcanic eruptions can also affect global temperatures. A particularly powerful explosion propels ash and other particles into the stratosphere, where they can remain for months. Sulfur released during eruptions may combine with water vapor to form droplets at high altitudes. Both might circle the globe, inhibiting incoming sunshine, which cools things down. However, the effect usually lasts for only a few months or years. Volcanic eruptions release greenhouse gases like carbon dioxide and methane into the atmosphere and can contribute to global warming. These gases sometimes seep out of the Earth between eruptions but their contribution to long-term changes in global temperatures is negligible.

Recorded Climate Shifts

The connection between rising temperatures and human carbon dioxide emissions was first suggested by Guy Callendar in 1938. Callendar analyzed data from 147 weather stations worldwide and discovered that average global temperatures had risen by 32.5 °F (0.3 °C) over the previous 50 years due to fossil fuel burning.

Two decades later, Charles Keeling decided to measure carbon dioxide levels in the air. Installing custom-made equipment at a weather station on Hawaii's Mauna Loa volcano, his observations revealed that the amount of carbon dioxide in the atmosphere was rising. The Keeling Curve, which documents changes in carbon dioxide levels over time, is the most extensive record of these emissions on the planet. They are still being monitored today.

The first accurate global climate model predicted in 1967 that if the amount of carbon dioxide in the atmosphere were to double, this would raise global temperatures by around 35 °F (2 °C).

John Mercer was working in Antarctica in 1968 when he found a freshwater lake in the Transantarctic Mountains, indicating that the West Antarctic ice sheet might once have melted away. He established that during the last interglacial period, temperatures were 42–45 °F (6–7 °C) warmer and sea levels 19.6 feet (6 meters) higher. He warned that this could happen again. His predictions came true in 1995 when parts of Antarctica's Larsen ice shelf began collapsing; nearly all of it was gone by 2017. Sea ice loss in the Amundsen Sea embayment is accelerating.

NASA's Nimbus III satellite, the first able to measure atmospheric temperatures, launched in 1969, was a game changer for climate science. For over 30 years, satellites have provided precise data on global climate and warming. They accurately measure land and sea surface temperatures, giving scientists a window on how Earth's climate is changing.

Ice cores drilled at the poles unlock the keys to Earth's climate history. The deeper they go, the older the record. A comparison of cores taken in 1985 and 1998 showed an unmistakable correlation between greenhouse gas emissions and higher atmospheric temperatures. In 2004, a 1.86 mile-long (3 kilometers) core drilled in the Antarctic provided 800,000 years of climate data. This revealed that carbon dioxide in the atmosphere was stable until the early 19th century, after which it began rising.

In 1998, the Intergovernmental Panel on Climate Change (IPCC) was launched. This organization reports to the United Nations and assesses climate change science. It has published five assessment reports, each confirming that the planet is warming due to human greenhouse gas emissions.

It was only in 2003 that scientists linked extreme weather events with climate change after thousands of Europeans died in a devastating heat wave. A scientific paper penned by a scientist at the UK Met Office the next year proved that climate change had doubled the risks of such an event occurring.

During the International Polar Year, which ran from March 2007–March 2008 and aimed to provide more scientific information on polar regions, researchers made some alarming discoveries. Ice sheets were melting at rates last seen over 10,000 years ago, with those in Antarctica and Greenland likely to disappear entirely. This would raise world sea levels by 33 ft (10 meters). Although sea levels were expected to rise gradually over hundreds of years, the scientists warned that this could happen much faster if global warming accelerated.

In addition, methane hydrate locked inside polar sea beds was degrading as temperatures rose. This would create methane plumes with the potential to accelerate climate change when the gas entered the atmosphere.

In 2021, climate scientists warned that the Earth was likely to reach 34.7 °F (1.5 °C) of warming far earlier than models had forecast and many adverse effects of climate change were now guaranteed, including:

- more intense rainfall spurring widespread flooding
- deeper and longer-lasting droughts
- sea level rise swamping coastal areas
- thawing permafrost
- ocean acidification

Their report stated that these changes could all be attributed to human fossil fuel burning. Catastrophic temperature increases would likely continue over the next two decades, breaching numerous climatic tipping points (UK Research and Innovation, 2023).

Modern Climate Anomalies

The future isn't what it used to be—and neither is the global climate. During the last 11,650 years, both the planet and humanity enjoyed one of the most stable climate epochs in history—the Holocene (Meyer & Newman, 2020). This enabled people to settle in favorable locations, start farming, and build communities and cities. Many scientists now believe that we are again entering a time of uncertainty caused by human actions. Some have dubbed this the Anthropocene—the age of man. One of the things we are inexorably changing is the climate. This has become glaringly apparent as the future approaches.

Short-Term Variations Versus Long-Term Trends

You might have come across these terms in reports about climate change. But what do they mean? Climate change is a long-term alteration of general climatic conditions, where temperatures have consistently trended upward over three decades or more (Wikipedia, 2013). However, there are highs and lows (anomalies) in the data over time because short-term natural events often raise or lower temperatures temporarily. These include volcanic eruptions, as well as the ENSO cycle in the Pacific Ocean.

The persistent warming of the global climate over the last 60 years is too intense to only be the result of natural variability. This is obvious when long-term trends are analyzed.

Changes in Ocean Currents and Consequences of THC Collapse

It is largely indisputable that climate change is a reality but exactly how things will play out is unclear. There is growing concern that continued warming will forever alter the climate system with serious consequences for the planet and humanity. One potential impact is the disruption of planetary ocean circulation, also known as thermohaline circulation (THC) or Meridional Overturning Circulation (MOC). This system acts like a giant conveyor belt, distributing heat and nutrients across the world's oceans.

Temperature and salinity, which affect seawater density, drive these giant circulatory mechanisms. The process begins at the equator, where winds push warm surface waters poleward. As the water moves toward the North Atlantic, evaporation increases its salinity and it cools down when it reaches the frigid northern zones. Both these factors raise its density, so it sinks at high latitudes. This dense water flows into ocean basins. Most of it wells up in Antarctica's Southern Ocean, where this now nutrient-enriched water forms the basis of the marine food chain, nourishing creatures from plankton to whales. Smaller amounts of water travel very slowly through the ocean depths, to well up in the South Pacific—this process takes around 1,000 years. Water from around the world intermingles in these deep ocean currents, transporting heat energy, dissolved solids, and gases.

THC influences the global climate. It may alter the effects of solar radiation, warm the polar regions, and regulate the formation of sea ice. Scientists are concerned that THC will be disrupted as the polar regions warm, releasing more freshwater into the ocean from melting ice. Because meltwater is less dense than seawater and is augmented by heavier rainfall events, there is less evaporation as the Atlantic MOC moves north. While the water will still sink and move southwards, because it is less dense it won't sink as deeply and will move sluggishly.

Research in Greenland appears to suggest that, prehistorically, melting ice reduced the MOC sufficiently to slow or even shut it down, which cooled the entire North Atlantic region. There is, however, some uncertainty as to exactly when this would occur; some scientists believe that there are already signs that the Atlantic MOC is beginning to slow down. Computer models indicate that the shutdown could occur at any time between 2025 and 2095 (Andrews, 2023). However, many of these predictions are based on temperature alone and scientists are working to understand more about the ocean's chemistry and how the MOC functions in order to make more reliable predictions.

In early 2023, researchers established that rapidly melting ice in the Antarctic could also disrupt deep ocean circulation, with flows potentially declining by as much as 40% by 2050 (Al Jazeera, 2023). While there have long been concerns about disruptions to the Atlantic MOC, it now seems that the slowing or collapse of the Antarctic circulation system could happen much faster. Besides reducing marine nutrient flows, this could also reduce the ocean's ability to store carbon dioxide.

Mounting Evidence

Climate deniers refuse to acknowledge that there is anything amiss with the global climate. However, there are many indications that climate change is occurring.

Satellite Data

Remote sensing and satellite data provide accurate information about the Earth's atmosphere, oceans, and temperature. Space agencies worldwide are using satellites to track greenhouse gas emissions, weather patterns, ice melt, bleaching corals, widening deserts, and changes in wildlife migrations. Satellites are also being used to monitor climate change drivers such as deforestation. The information they provide is accurate enough to inform scientific predictions.

Arctic and Antarctic Ice Melt

A 2021 study found that ice melt worldwide is accelerating with the Greenland and Antarctic ice sheets leading the way. This unfortunately corroborated the IPCC's worst-case scenario. About 28 trillion tons of ice were lost in these areas between 1994 and 2017. While atmospheric warming was the main culprit, ocean warming also played a significant role. From the early 1990s until 2017, the rate of melt, particularly from floating ice, increased by as much as 57% (Harvey, 2021).

Disappearing Glaciers

In the same study, it was found that glaciers had also reduced considerably, losing over six trillion tons of ice between 1994 and 2017—nearly 25% of all ocean ice loss. Diminishing glaciers pose several threats to humanity, including reduced freshwater supplies, flooding, and impaired agricultural production (Harvey, 2021).

A 2023 study suggested that all the world's glaciers would probably disappear by the end of the 21st century. Glaciers at the poles often extend out over the ocean and are affected by warming waters, as well as warmer atmospheric temperatures. Even if warming were to stabilize at 34.7 °F (1.5 °C), about half the world's glaciers would be lost, raising global sea levels by at least four inches. Alaskan glaciers, together with smaller ones around the world, are most likely to contribute to sea level rise (Mooney, 2023).

Coral Bleaching

Bleached corals lose their vivid colors and turn into pale skeletons on the seabed. This happens because these colonies of minute animals depend on a particular algae to survive. These provide the coral with oxygen and food, while the reefs shelter the algae. It's these algae that give coral it's amazing colors. For an organism that forms huge reefs, coral is surprisingly delicate. It is sensitive to water pollution, changes in water chemistry, and significant temperature increases. As the water warms, they expel their algal friends in a stress response. This means that the coral turns pale and bleaches. Without the algae, the coral eventually dies. If conditions return to normal quickly enough, the coral can recover from the setback.

Besides hosting an incredible array of diverse marine life, corals have many benefits for humans. Their reefs provide seafood, protect coastlines from erosion and storms, and absorb and store carbon dioxide. Between 2014 and 2017, 75% of coral reefs worldwide experienced sufficient heat stress to cause bleaching.

Coral died on 30% of these reefs. As sea temperatures increase, so do the risks of more bleaching events. Some sources believe that if this continues, there may be no more coral reefs left on Earth by the end of the 21st century (Hancock, 2023a).

Key Chapter Takeaways

- In the second decade of the 21st century, there's no denying that climate change exists. Your town or city is probably experiencing hotter, warmer weather than before. Perhaps you've survived an extreme weather event.
- The greenhouse effect is a natural phenomenon where warming gases in the atmosphere, such as carbon dioxide, nitrous oxide, and methane, together with water vapor, prevent solar radiation from escaping back into space. This keeps Earth warm enough to support life.
- However, since the 1940s, there have been consistent indications that the climate is warming, mainly due to humanity's insistence on burning fossil fuels. There is now ample evidence that climate change is happening and things are getting warmer at the Earth's surface. Ice caps and glaciers are melting, forests and homes are burning in wildfires, hurricanes are roaring along coastlines, sea levels are rising, and corals are bleaching. Floods and droughts are other consequences of global warming.

Climate change and the failure to take meaningful action are starting to cost us dearly. More billion-dollar disasters mean massive government spending on disaster relief, relocations, and rebuilding. That means less money for other things. Livelihoods

are lost and regional economies suffer. People struggle with phys-
ical injuries, PTSD, and missed opportunities. Climate change is
starting to take its toll on individuals, businesses, and even entire
nations. We'll explore these costs and their implications in the
next chapter.

Billion-Dollar Disasters—A Deeper Dive

Since 1980, the US has experienced 341 climate- and weather-related disasters resulting in over $1 billion in damages, with the cumulative costs totaling $2.475 trillion. Between 1980 and 2000, some 75% of damages caused by billion-dollar disasters were due to climate change; by 2022, this had increased by 10%. Between 2016 and 2021, there were 122 billion-dollar disasters which killed 5,000 people. According to the NOAA, the escalation of extreme weather events looks set to become the new normal, with landfalling category 4 and 5 hurricanes contributing significantly to damage costs (Smith, 2023).

These are conservative estimates, however, as they only account for billion-dollar extreme weather events. Numerous multi-million dollar events occur throughout the US every year, the costs of which can be added to this tally.

Climate Change Costs in the US

Several states in the Central, South, and Southeastern US, together with the Caribbean, are particularly vulnerable to—and impacted by—severe weather events fueled by climate change. The following states have incurred the most cumulative damages in billion-dollar disasters since 1980:

Ranking	State	Cumulative costs of billion-dollar disasters since 1980
1	Texas	$380 billion
2	Florida	$370 billion
3	Louisiana	$290 billion
4	California	$135 billion
5	North Carolina	$81.6 billion
6	Mississippi	$78.5 billion
7	New York	$75.4 billion
8	New Jersey	$59 billion
9	Iowa	$55 billion
10	Missouri	$47.5 billion
11	Illinois	$46.9 billion
12	Alabama	$46.8 billion
13	Georgia	$38.7 billion
14	Colorado	$37.4 billion
15	Tennessee	$36.9 billion
16	Oklahoma	$35.6 billion
17	South Carolina	$33.5 billion
18	Kansas	$32.2 billion
19	North Dakota	$31 billion

(Smith, 2023; USA Facts, 2023.)

Deaths Per Year from Billion-Dollar Disasters

The following table details the number of deaths per year from climate-related severe weather events from 1980–2022:

Disaster Event Type	Deaths Per Year
Tropical Cyclone	160
Heatwave and Drought	99
Severe Storm	46
Winter Storm	33
Flooding	16
Wildfire	10
Freeze	4
Total	368

(Climate.gov, 2023.)

Hurricanes, Floods, and Wildfires

As you can see from the above tables, some of the costliest natural disasters in terms of human lives and damages have been the result of these three events. So how does climate change escalate the effects of severe weather?

Increased Frequency and Intensity

There are several ways in which global warming amplifies extreme weather events. Heat waves worsen because there are more extremely hot days and nights in summer. The heat increases inland evaporation rates, which may deepen droughts and raise the likelihood of wildfires. Fire seasons are also longer and more severe. As mentioned previously, the presence of more water vapor in the atmosphere leads to heavier rainfall and bigger snowstorms. El Nino events, which reduce rainfall in many areas, can intensify droughts, while La Nina conditions favor heavier precipitation. These natural climate variations are likely to worsen due to climate change.

The heightened levels of heat and moisture in the atmosphere create larger, more powerful hurricanes that last longer, produce more rainfall, and could make landfall in new places. Rising sea levels mean that the high tides and giant waves generated by storm surges are occurring closer to coastal infrastructure and habitation. The flooding these events generate is likely to worsen, with more communities being affected.

How Urbanization Impacts Disaster Risk

More than half the world's population has settled in urban areas and it is here that most economic and financial hubs are located. Urban migration is exploding and it is estimated that two-thirds of humanity will live in cities by 2050 (CDP, n.d.). Resources like water, food, and energy need to be brought into these centers and waste needs to be removed. Urbanization also impacts rural communities, some of which might supply urban needs, although many rural people migrate to cities hoping for a better life.

Cities once provided economic opportunities and refuge during tough times. However, this is no longer the case. According to the Carbon Disclosure Project (CDP), the top five climate-related hazards cities are most likely to experience include:

1. surface and flash flooding
2. heatwaves
3. rainstorms
4. extremely hot days
5. droughts

Research indicates that, if nothing is done to address climate-related threats, significantly more urban dwellers may be affected by heat waves, while some 800 million people could be exposed to rising seas and storm surges (CDP, n.d.). Cities tend to focus on short-term hazards while downplaying or ignoring medium- and long-term risks, potentially imperiling citizens.

Social services and infrastructure are already strained due to the rapidity of urban migration. Climate change will place more burdens on cities while disrupting economic operations and government services. Another ticking time bomb is the threat of deteriorating public health under climate change, with heat, droughts, and vector-borne diseases increasing dramatically, especially under a 35 °F (2 °C) warming scenario. These challenges will disproportionately affect the poorest, most vulnerable residents.

While many cities are well aware of the risks, little is being done to address them. For African and Latin American cities, the primary hindrance is a lack of access to basic services, while North American and European cities are challenged by poor infrastructure and budget constraints.

Creating More Resilient Cities

Cities are often confronted by multiple, simultaneously occurring challenges and stresses, such as recurrent flooding, limited social relief, and significant unemployment, which may weaken communities over time. Acute shocks such as hurricanes or earthquakes are then amplified.

Urban resilience means that all aspects of a city, from its systems to its businesses and communities, can survive, adapt, and thrive regardless of adverse events. Cities need to evaluate all their capacities and risks while engaging with their most vulnerable citizens. Access to quality, accurate information is essential to enable resilient cities. Urban enclaves are not silos, where separate teams work on different aspects of governance, planning, and disaster management; they consist of people and places, and all sectors need to work together to achieve resilience.

Some cities are already working to mitigate disaster impacts and ameliorate climate change risks. The most common actions cities are taking include flood mapping, long-term planning, crisis management, community engagement, and tree planting.

Natural Barriers—Mangroves and Wetlands

Mangroves are a rare ecosystem that usually thrives in places where land and sea meet. These ecosystems are biodiverse and important nurseries for fish and shellfish. This makes them important revenue generators for surrounding communities. Besides all this, mangroves naturally protect coastlines. They have large prop roots that anchor them in the soil, enabling them to withstand the ravages of tides, winds, and storms. They are accordingly a buffer against rising sea levels, storm surges, and tsunamis. Their roots filter water and trap sediment, thereby slowing erosion and stabilizing shorelines.

Using mangroves for coastal defense is thought to be five times more cost-effective than manmade solutions. They also help to combat climate change by taking up atmospheric carbon and storing it in their roots. They may absorb four times as much carbon as a rainforest of similar size (Montague, 2021). This remains locked inside the trees when they die, as they usually slide underwater.

Several countries are using these natural, if threatened, trees to protect their coastlines. In the US, mangroves are being established on the coast of Florida, a state that has been hard hit by the devastating weather climate change is bringing to its shores.

However, mangroves themselves face new threats in the age of climate change—they may become overwhelmed by rising seas and are sensitive to water temperature changes.

Climate Risks and Insurance Costs

Climate change is creating several economic risks. Insurance companies are feeling the pinch and many have gone bankrupt after receiving an avalanche of claims in the wake of wildfires, hurricanes, and floods. Unmitigated climate change impacts could contract regional economies as well as the global economy. Not only that, the UN Climate Finance Committee recently established that climate-related funding is inadequate and is being facilitated by loans, which may make it unsustainable (UNFCCC, 2023).

Climate Risk Insurance—A Two-Edged Sword?

Climate change is creating unexpected risks for insurers and company presidents contend that severe weather events are destabilizing the industry. In 2021, as many as 60% of insurance risk managers held that climate change would make insurance coverage in certain locations unworkable (Ross, 2021).

Insurers now need to peer into the future to determine their risks rather than considering past scenarios. The 2017 and 2018 Californian wildfires, which were probably amplified by climate change, evaporated 25 years of profits for local insurance companies. Globally, 72% of insured losses between 2016 and 2018 were caused by wildfires. In the Southeastern US, hurricane losses and litigation costs in Louisiana and Florida bankrupted several insurers. These states were obliged to borrow money to ensure that all claims were paid (Hill, 2023).

In many parts of the world, climate stress testing is becoming the norm and results have concerned insurers. Tests conducted in France in 2021 established that insurance risks could increase as much as five-fold in disaster-prone regions. This would potentially raise premiums by 200% within 30 years (Ross, 2021).

Insurance companies have adopted various approaches. Some have raised premium prices, making insurance less affordable. Starting in January 2022, 31 US states saw home insurance prices skyrocket by as much as 20–30%. This means that many homeowners can no longer afford property insurance (Hill, 2023).

Other insurers have simply refused to insure certain homeowners, reducing or limiting coverage for homes on the east coast that might be at risk from flooding, and those on the west coast that could be subject to wildfires. In Louisiana, insurance companies are no longer covering hurricane-prone areas.

Individuals unable to obtain insurance are turning to state-backed insurers to fill the gap. In California, many have opted for the state's FAIR plan, an association of insurers that are a provider of last resort. Between 2018 and 2019, such policies rose by 36% state-wide, while the number of policies held in Louisiana's state-run programs tripled. Those who choose to remain uninsured will be forced to rely on FEMA and other government relief when disaster strikes (Hill, 2023).

Climate Change and the Global Economy

Climate change may not only depress local and regional economies in places subject to extreme weather, but is also likely to impact the global economy. Swiss researchers estimated in 2021 that climate change impacts could make the global economy 10% smaller by 2050 if temperature increases remained on the current trajectory. However, if temperatures increased by 37.7 °F (3.2 °C), this could rise to 18%.

Asian countries are likely to be hardest hit, with Gross Domestic Product (GDP) expected to contract by 5.5% in a best-case scenario and 26.5% in a worst-case one. The Chinese economy, one of the world's biggest, could reduce by as much as 24% in a worst-case scenario. Middle Eastern and African economies would contract by 4.7% if temperature increases stayed below 35 °F (2 °C) and as much as 27.6% in a worst-case scenario (Marchant, 2021).

Recovery and Investment Needs

Although around 27% of investments in recent years have been channeled into climate change adaptation to avoid future loss and damage, financial flows to developing countries are nevertheless estimated to be at least 5–10 times below 2050 requirements. Complicating matters, it is often difficult to separate this funding from development assistance, especially as much of the latter automatically includes climate-proofing.

Loans from multinational development banks provide most climate-related finance—some 72% of public climate finance between 2016 and 2020 was obtained through loans, while grants accounted for 26%. Using loans to finance climate mitigation and adaptation measures is not necessarily sustainable as it may lead to high debt levels for donor countries. In addition, there are generally insufficient funds to provide adequate humanitarian aid in the wake of natural disasters sparked by climate change. In

2020, only 26% of all humanitarian aid was allocated to climate-related emergencies.

Developing countries have very little risk insurance coverage; gaps are as high as 97% in some cases. Initiatives have been developed to increase the uptake of insurance by both public entities and private households (UNFCCC, 2023).

Vulnerable Populations

Some nations and communities will be disproportionately encumbered with humanity's climate change burden. Many are developing countries, which have historically had very low emissions. Small island nations have contributed less than 1% to global greenhouse gas emissions, for example (Wikipedia Contributors, 2023b).

Small Island Nations

Small island developing states, such as those in the Caribbean, Pacific, and Maldives, are particularly vulnerable to climate change and are already experiencing significant impacts. Because of their low profile in the ocean and small land masses, island nations have considerable exposure to the extreme weather climate change delivers.

Dangerously powerful hurricanes and a rise in sea levels present major threats. As you might expect, the latter is the greatest fear for small island nations, as it reduces the size of habitable land and threatens traditional cultures. Food security may be affected, as saltwater inundations could make what little arable land is available impossible to cultivate. If climate change continues unabated, many of these will literally drown. Some—but not all —islands surrounded by reefs might adapt to changing circumstances, although those with more development could experience significant impacts.

Other threats include coral bleaching, reduced freshwater supplies, and alien species invasions. Due to sea level rise, flooding, and extreme storms, island nations may face several health risks, including contaminated freshwater, droughts, and diseases spread by mosquitoes and other insects. Ocean acidification may alter the distribution and survival of marine food sources. Climate change is placing tourism, an important revenue generator for many, at risk.

These nations have few resources to protect their islands from the onslaught and many are relying on international aid to deal with severe weather events. Inhabitants would like to leave but have no resources, so there is considerable pressure on governments and community leaders to implement adaptation measures. The Asian Development Bank has committed significant financial resources to island nations to assist them in implementing disaster risk management, nature-based solutions, and gray-green infrastructure combinations.

Many island nations spend considerable amounts on importing fossil fuels, although many are starting to change their economic models. Barbados, for example, is developing its renewable energy sector and has rolled out over 50,000 solar-water-heater installations (Wikipedia Contributors, 2023b). Island nations have turned to advocacy in the international community to encourage climate change mitigation and adaptation.

Poverty and Inequality

Thirty years of research has proved that it is usually the poor that bear the heaviest burden when natural disasters strike. They often reside in hazardous areas and lack the funds to mitigate their risks. Disasters cost them their livelihoods, health, and food security, not to mention their lives. Climate change could derail the UN Sustainable Development Goals, one of which is to eradicate poverty by 2030 (UNDRR, 2021). While losses tend to be higher for wealthier citizens, low-income groups experience more diverse severe impacts.

Most of the urban poor are informally employed and often use their homes to generate income. Losing houses and possessions in natural disasters is significant, especially in the absence of formal employment, insurance, and other safety nets.

Rural livelihoods are particularly vulnerable to weather-related disasters and many people have little capacity or capital to recover from agricultural or infrastructure losses.

Inequality means that risks are transferred from those who benefit from their cause to those who carry the cost. Uneven economic development, resource overconsumption, segregated development, and climate change foster inequality. There are concerns that rising inequality could destabilize many parts of the globe, hampering attempts to manage natural disaster risks.

Climate Justice—Unequal Impacts and Responsibilities

The concept behind climate justice is that those who have emitted copious greenhouse gases to amass wealth should assist those negatively affected by climate change, especially the most vulnerable countries and communities. Countries with lower per capita income generally contribute the least to global greenhouse gas emissions. Equality and human rights should inform all climate change actions.

The impacts of climate change may be more severe for some groups than others, even in the same country. Women, the disabled, and indigenous peoples are more severely affected by climate change, for example, and often do not have access to resources to reduce their risks. Low-income countries and vulnerable groups within them are more susceptible to damages and losses caused by climate change.

Children and young people who have not contributed to the climate crisis will not be able to escape its effects. The decisions of previous generations have impacted their human rights and these need to be considered when climate change decisions are made and action taken.

Climate change is as much a human rights issue as it is an environmental one. People are losing their lives, income and livelihoods, cultures, and access to food and water. They are experiencing air and water pollution, extreme weather events, displacement, and conflict. The health impacts of climate change include disease outbreaks, heat stress, malnutrition, and PTSD. Vulnerable populations experience these risks more severely and have fewer resources to mitigate them. Deaths from droughts, floods, and storms were found to be 15% higher in vulnerable regions than in less vulnerable ones between 2010 and 2020 (UNDP, 2023).

Achieving climate justice is not easy. Some obstacles include:

- Women, young people, marginalized groups, and indigenous people are often excluded from climate negotiations and mitigation plans. There is also a general lack of transparency.
- People, often those most affected, generally have little access to relevant resources and education. This prevents

people from understanding their rights and participating in decision-making.
- Environmental activists and defenders who demand environmental rights and climate justice may be jailed, threatened, abducted, or even murdered.
- There is insufficient financial and technical support from wealthier nations. (UNDP, 2023.)

Efforts are being made to improve climate justice. Vulnerable communities are now being acknowledged during negotiations. Activists and young climate leaders have taken to the streets and raised awareness of climate impacts and injustice. In 2022, the UN General Assembly declared that access to a healthy, clean, and sustainable environment is a human right for people everywhere. In 2023, the organization requested the International Court of Justice to establish different countries' climate obligations. New funding was established at COP27 (the 27th United Nations Climate Change Conference of the Parties) to help vulnerable countries and communities mitigate climate-related loss and damages (UNDP, 2023).

Recovery and Resilience

There are several ways in which countries, regions, and communities can recover from natural disasters caused by climate change and build resilience to help avoid future damage.

The Role of International Aid

As climate change intensifies, communities are being affected by consecutive disasters, with little time to recover before the next one arrives. The most vulnerable people could fall through the cracks if their needs are not heard and understood.

According to the International Federation of the Red Cross (IFRC), climate change and its associated weather disasters could leave 200 million people a year needing humanitarian aid by 2050. This is expected to cost around $29 billion annually. However, the stimulus packages launched following COVID-19 indicate that funding resources are available for climate change adaptation and disaster management. For example, it would take around $50 billion to meet the adaptation requirements of 50 developing countries for the next ten years (IFRC, 2020 & 2022).

It is important to ensure that financial resources are properly allocated so the aid reaches those who need it most. It often appears that the most at-risk countries with the lowest capacity to adapt are not necessarily a priority for funding allocations. An analysis conducted by the IFRC in 2020 revealed that none of the 20 countries most vulnerable to climate change were among the 20 top funding recipients. Somalia, one of the hardest hit, was only listed 71st for funding disbursements, for example. Only 38 out of 60 vulnerable countries and five out of eight highly vulnerable countries received funding—and were allocated the equivalent of $1 per person.

Even if funding does reach these countries, it is estimated that only 10% reaches locals on the ground, as donors tend to focus on national infrastructure projects. Donors don't always track how their funds are spent and are unable to assess the degree to which their aid reaches the most vulnerable communities.

In 2016, developed countries signing the Paris Agreement agreed to allocate at least $100 billion a year to developing countries until 2025 to facilitate climate change adaptation and disaster mitigation. However, funds received have fallen significantly short of this (IFRC, 2022).

Smart financing is when funding strategies combine the right types of funding suitable for a specific situation. Many efforts to adapt to climate change and mitigate its effects need to address interconnected needs, yet donors often treat these needs as separate problems. Understanding the big picture is essential to identify and address funding gaps. As the global economy becomes less resilient, funding needs to be efficient, coherent, and pertinent.

Smart financing brings together different donors to provide investment in disaster risk insurance, prevention, social protection and relief, and rehabilitation. This layered funding approach needs to be strategic and appropriate. Participatory financing is more likely to address the real risks and impacts faced by communities.

Adaptation to Climate Change and Disaster Risk Reduction

Disaster risk reduction (DRR) and climate change adaptation (CCA) often occur concurrently.

DRR aims to reduce disaster risks by:

- analyzing and managing their causes
- reducing exposure to hazardous events
- reducing the vulnerability of people and their belongings
- improving environmental and land management
- preparation for adverse events

CCA usually involves the following:

- adjusting economic, social, or ecological systems to respond to actual or expected climate change events and their impacts
- adapting development to changes in average precipitation, sea levels, and temperatures
- managing and reducing the risks posed by escalating extreme weather events (UN Food and Agricultural Organization, 2016)

Preparing for Disasters

Even if the world were to cease emitting greenhouse gases with immediate effect, certain climate impacts are now locked in. This means that countries are having to consider both CCA and DRR. However, implementing these can be challenging.

Before either of these actions can take place, it's essential to have high-quality risk assessments and projections to address local issues and consider the national situation. This enables policymakers to make informed decisions. Yet many countries either do not have this information or are relying on outdated data. The UN Development Program assists countries with vulnerability assessments, risk projections, and DRR plans.

Nationally Determined Contributions (NDCs) and CCA should complement one another, with CCA reinforcing NDCs. Several projects are helping countries formulate National Adaptation Plans while revising their NDCs and securing CCA funding and investment.

As mentioned previously, local concerns are paramount when considering and implementing adaptation and DRR measures. The Global Commission on Adaptation is helping to ensure that locally led and owned approaches to adaptation are both sustainable and effective.

Emission reductions bring benefits like cleaner air, improved health, and better livelihoods. Experiencing these advantages spurs climate action. Progress is accelerated due to the linkages between climate action and sustainable development.

Private sector engagement can help with adaptation. Governments can work with chambers of commerce and other business groups. Over 60 countries are engaging with their private sectors to implement or enhance their NDCs (UN, n.d. All About the NDCs).

Budgeting is essential—plans for which there are no budget will not come to fruition. Countries need to integrate CCA into national planning and budgeting processes.

Countries need to monitor their progress to establish whether their mitigation and adaptation plans are practical and resources are being allocated efficiently. Monitoring is important for reporting purposes.

Sustainable Rebuilding Efforts

Disaster recovery refers to the return to normal after a crisis. It involves resolving all the different impacts of the event. However, the rebuild should improve on previous development and increase future resilience. Sustainability should therefore feature in disaster recovery plans. This can also help to address poverty and inequality.

Sustainability refers to the ability to provide for present needs without compromising those of future generations. When it comes to disaster recovery, sustainability can reduce the impacts of future disasters by rebuilding so that infrastructure and ecological systems are more resilient. This improves the long-term quality of life for affected communities. Local stakeholders can be empowered by participating in the rebuild, especially if they are from vulnerable groups. Some options for rebuilding sustainably include green buildings, restoring ecosystems, conserving natural resources in planning and construction, finding ways for the community to participate in the rebuild, and strengthening government and organizational capacity.

Sustainability can be aligned with national and regional objectives. It is important to ensure that the social, economic, and environmental dimensions of disaster recovery are considered. If there is a lack of scientific support for a particular action, it should be avoided or reduced. The best available knowledge, including indigenous knowledge, should be used when planning and implementing disaster recovery. This will ensure a positive outcome from the crisis.

Key Chapter Takeaways

- In this chapter, you have learned how the frequency and severity of natural disasters related to severe weather are starting to take a significant financial toll on countries across the globe and may even affect the global economy.
- Things are becoming so serious in some parts of the US that large insurance companies will no longer provide insurance for properties considered to be high risk due to their locations.
- Economic recovery after disasters is generally slow, as investors are reluctant to reinvest, while business and

property owners sometimes have insufficient funds to bounce back quickly.

- Extreme weather events and natural disasters fueled by climate change are disproportionately affecting society's most vulnerable communities in both cities and rural areas. There have been few attempts to create more resilient cities that are capable of recovering rapidly from weather-related disasters. Small island nations and many developing countries, which have historically contributed little to climate change, are already seeing some of the worst impacts.
- Climate justice theorizes that wealthy nations who are responsible for significant historic emissions from fossil fuel burning and industry should help to fund initiatives and investments to relieve the burden and help such nations and people adapt to a future under climate change.
- Climate change funding is generally considered to be insufficient and well below what will be needed to assist the worst-affected nations by 2050. Much climate finance is also funded by loans from international development banks, this funding could be unsustainable in the long run due to the debt risks it poses to donor countries. Smart funding, which considers all impacts of disasters and climate mitigation measures, could be the answer.

You might be wondering how a tropical storm can develop into an intense category 5 hurricane that sets off a significant natural disaster. How does drought create damaging winds capable of fueling a wildfire that burns for weeks? In the next chapter, we'll take a look at the giant weather systems causing billion-dollar disasters across the globe.

Understanding Shifting Weather Patterns

Have you ever wondered why it has snowed in places that are usually warm and sunny— or why some regions experience incessant floods, while others struggle with constant drought? In this chapter, we will delve into the intricate dance of weather and climate, unraveling the factors that create and intensify extreme weather across the globe.

Weather Versus Climate

With the advent of climate change, the difference between weather and climate can be confusing. We hear of severe weather, as well as how global warming is influencing climate change. So, what is the difference between weather and climate?

Weather is a short-term forecast determined on a daily basis that lasts for just a few days. As we all know, weather changes throughout the day and across the seasons. The weather forecast helps us to plan our days. Can we mow the lawn this afternoon? Should we take an umbrella when we go out?

In contrast, climate refers to long-term weather patterns—the average daily weather over an extended period in a specific place. You know you can expect snowfalls in the US Northeast in January and humid, overcast conditions in July in the Southeast, for instance (NOAA, 2009, Oct 2). Long-term climate records also include unusual weather, such as particularly hot or cold days, or unexpectedly high precipitation. Climate change is constantly altering long-term climate records.

Daily Fluctuations and Long-Term Averages

All this doesn't mean that the climate never changes. Climate variability refers to times when the climate temporarily deviates from the average. This may be caused by changes in ocean and air circulation or even volcanic eruptions.

Global annual temperature fluctuations also occur, so some years may be warmer or wetter than others. When long-term trends are analyzed, however, these trends show temperatures rising overall due to climate change. Climate variability increases the chance of severe weather.

The Role of the Atmosphere

The atmosphere is a gaseous layer surrounding the Earth. It makes air breathable and life possible. The atmosphere extends about 6,214 miles (10,000 kilometers) above the Earth's surface. Rather like a layer cake, it consists of the following strata, each of which has a layer or pause between them:

- **Troposphere:** Closest to the Earth's surface is the troposphere, which extends about 9.3 miles (15 kilometers) into space. It is thickest at the equator and thinnest at the poles, with temperatures that become colder with altitude. The troposphere is where clouds form and weather events are born, as it contains the most water vapor. Recent research indicates that the troposphere is heating and expanding but scientists believe this will not have a significant effect on climate change (Kreier, 2021).

- **Stratosphere:** The stratosphere is much thinner than the troposphere and contains ozone. At ground level, ozone is a pollutant but, 9.3–31 miles (15–50 kilometers) up in the stratosphere, it protects the Earth from potentially harmful UV rays.

- **Mesosphere:** The mesosphere extends 31–53 miles (50–85 kilometers) from the Earth's surface. Temperatures here can get as low as -130 °F (-90 °C). The mesosphere develops noctilucent clouds, silvery or blue-colored clouds visible at night. These are the atmosphere's highest clouds.

- **Thermosphere:** The final, extremely thin layer is the thermosphere, which is located 373 miles (600 kilometers) above Earth's surface. During the day, solar radiation raises temperatures here to a red-hot 3,623 °F (2,000 °C) (Vaia, n.d.).

Atmospheric gases consist mainly of nitrogen, followed by oxygen, argon, and others like carbon dioxide and water vapor.

Atmospheric pressure changes are caused by changes in air density, which are temperature-related. These pressure changes are measured on barometers and predict weather changes. A rapid drop in pressure signals the arrival of a low-pressure system associated with cloudy, wet, and windy weather. Rapid increases in atmospheric pressure indicate the presence of a high-pressure system. High pressure pushes the clouds away, creating sunny, hot, and dry weather.

Atmospheric Air Circulation

Just as ocean water circulates, air masses also circulate in the atmosphere. The bottom of the troposphere warms up, especially near the equator and the tropics. This makes the air lighter, and the warm air rises. What goes up must come down, so the air drifts toward the poles, losing heat as it goes. As it cools, its density increases and it sinks. This creates a high-pressure region near the poles, balancing the low pressure near the equator. This drives the wind.

Winds blow from mid to low latitudes and from the tropics to the poles. Their movement is complicated by the Earth's spherical shape and spinning motion, which makes equatorial winds stronger at the equator while creating calm conditions at the poles. The westward winds blowing toward the equator are the trade winds. Closer to the poles, winds moving from west to east —the Westerlies—collect heat at the equator and move it toward the poles. This creates the polar vortex weather system.

Unpredictable Events and Extremes

Globally, current weather patterns are breaking decades of records, delivering intense heat waves, more severe droughts, and damaging storms. For weathermen and climate scientists, these extremes are a sinister warning.

Record-breaking Temperature Events

Climate change is spawning record-breaking heatwaves across the Northern Hemisphere, with every summer more scorching than the last. However, plummeting winter temperatures and frigid air may also be caused by a warming climate.

Heat Waves

Europeans plunged into every available fountain, lake, or stream in a desperate attempt to cool off during the summer of 2013. Temperatures soared, together with warnings to stay indoors and ensure that pets were hydrated.

That summer smashed heat records across the Northern Hemisphere, especially in July. Unprecedented heat waves scorched the Southwestern US, Mexico, Southern Europe, and China. The hottest-ever daily minimum temperatures were experienced in numerous places, including Death Valley, California, in the US, where daytime temperatures reached a staggering 130 °F (54 °C).

Innumerable people were hospitalized for heat-related ailments. Severe heat waves caused countless fatalities. Accurate record-keeping is not always a priority, so it's challenging to determine the exact number of heatwave-related deaths.

As forecast by the IPCC, heat waves are fast becoming part of the new normal, with China, North America, and Europe seeing more frequent heat waves in recent years. Without global warming, such events would only occur in China once in 250 years and almost never in North America, Mexico, or Southern Europe. Thanks to climate change, record-breaking heat waves are anticipated once every 15 years in the US and Mexico, once every ten years in Southern Europe, and once every five years in China. Heat action plans are being implemented to lower heat-related mortality and cool down cities through initiatives like tree planting (World Weather Attribution, 2023).

If humans persist in burning fossil fuels, heatwaves will become more common and last longer. Heat waves like those that occurred in the 2023 summer could take place every 2–5 years if global temperatures increase by 35 °F (2 °C) or more.

Record-Breaking Cold Snaps

The 2020–21 Northern Hemisphere winter saw two record-breaking incidents of frigid air in China, where Beijing and Tianjin experienced their lowest temperatures in 54 years. Temperatures plummeted to just above -0.3 °F (-19 °C) in both cases. Temperatures also dropped to all-time lows in the North American Midwest and Deep South, where Houston, Texas, recorded historic lows of 1.7 °F (8.3 °C) (Lin, 2021).

Chinese researchers attempted to unravel the riddle. They found that new warming combined with unusual jet stream and polar vortex conditions. Aided by the jet stream—air currents that move west to east—the polar vortex brings cooler winter weather to the Northern Hemisphere. Sudden temperature increases in the stratosphere can dramatically change the behavior of both the polar vortex and jet stream. Ocean temperatures and changes in Arctic sea ice also played a role, with unusual warming occurring in three ocean basins simultaneously (Lin, 2021).

Several factors may contribute to extreme weather events—and may occur over such wide geographical areas that they appear to have little connection to one another. A severe weather event may be driven by global warming, melting polar ice, temporary changes in ocean temperatures, and atmospheric circulation.

Unseasonal Weather

This term refers to haphazard weather patterns. Blistering heat follows frigid cold. Spring begins with record high temperatures, or summer heat wave conditions persist into winter. A good example is the conditions experienced in many parts of the eastern US in the 2022–23 winter season.

In late December 2022, an Arctic blast drove unusually cold air across the eastern US. Across much of the Midwest, a bomb cyclone developed along the cold front, bringing ferocious winds and blizzards. According to NOAA, a bomb cyclone is a fast-developing storm that occurs when atmospheric pressure drops at least 24 millibars over a 24-hour period. Lake-effect snow in New York and Canada intensified the storm's effects, dumping 50 inches (125 centimeters) of snow in five days (Voiland, 2023). In the Southeast, the icy air birthed tornados and put power and water supply systems under pressure. Christmas 2022 was a traveler's nightmare.

In the early days of January 2023, the weather did an about-face, as conditions abruptly shifted from unseasonably cold to unusually warm; temperatures rose around 20 °F to 30 °F higher than usual for the eastern US at that time of year. The dramatic spring weather broke seasonal temperature records across many central and southern states. Record-breaking warmth also brought an abrupt end to an equally icy winter across Europe as warm air barreled in from West Africa.

A forecaster explained the strange events. A weakening polar vortex in the stratosphere brought Arctic air into the mid-latitudes. These effects are short lived; when conditions return to normal, temperatures rapidly increase. Disturbances to the polar vortex have increased since 1980. In 2021, scientists found that declining Arctic sea ice in the Barents and Kara Seas resulted in higher snowfall events in Siberia, together with polar vortex disruptions and mid-latitude cold. Both sea ice declines and heavier snowfalls are linked to climate change (Voiland, 2023).

Marine Ecosystems

Like the atmosphere, the ocean also experiences extreme conditions linked to climate change. Marine heatwaves, dead zones, and coral bleaching are a few examples—and are expected to intensify. Besides acidification, climate change reduces the oxygen present in seawater, as warmer water holds less oxygen.

This has consequences for people and nature. As ocean temperatures rise, fish and marine life are moving to cooler areas. This not only disrupts ecosystems, it creates confusion in fisheries, and uncertainty as to regulations. Some species are deserting traditional feeding grounds, so fishing fleets need to travel further, which adds to costs. Changing water temperatures and chemical conditions have altered seasonal events like spawning, so some species' numbers peak earlier or later than before.

Marine heatwaves can bring both expected and unexpected environmental change. Between 2018 and 2021 after a period of historically high crab abundance and a series of marine heatwaves the population of snow crab in the Bering Sea declined by ten billion. (Szuwalski et al. 2023) The increase in water temperature caused an increase in the crabs' caloric needs which led to an unexpected mass starvation event.

All these changes may lead to diminished catches and revenues, increasing consumer prices.

Agricultural Impacts

US crop agriculture is essential for the local and global food supply, as US farms supply 25% of the world's grain. As of 2022, the ten states with the highest agricultural output in the US were

California, Illinois, Indiana, Iowa, Kansas, Minnesota, Nebraska, North Carolina, Texas, and Wisconsin (ERS, 2023).

Crop yields may be affected by changes in temperature and carbon dioxide levels, together with extreme weather.

Even more farmers worldwide are listing climate change as the top threat to production and livelihoods. The agricultural sector is particularly sensitive to changes in temperature and weather patterns. There are several ways in which climate change is detrimental to agriculture, potentially affecting food supplies and rural economies.

Changing Growing Seasons

All land managers are seeing seasonal changes due to climate shifts. These include that:

- Growing seasons have lengthened and temperatures have increased. This could create opportunities to grow new crops or crop varieties for different markets. However, this could also result in agricultural pests producing more offspring per season, potentially reducing pesticide efficacy.
- Changing rainfall patterns have altered water availability, while torrential rains and floods have become more frequent. This could deliver surplus water during the off-season, with less water available during critical growth periods. More intensive water management will be required to ensure that there is sufficient water at crucial times.
- Earlier spring thaws and later first frosts could enable more productivity, if temperatures allow adequate growth, there are sufficient nutrients, and diseases and

pathogens are limited. However, should winter
dormancy end too soon, fruit trees may develop buds
too early, becoming susceptible to late frosts.
- Early snowpack melt could reduce the quantity of
runoff water during prime irrigation periods in areas
where farmers use meltwater for crop irrigation.
- Higher summer temperatures often increase wildfire
threats, which may destroy grazing and croplands.
- Grain storage methods may need to be altered due to
winter temperature and humidity changes.
- Farmers may need to monitor their soils more carefully
to prevent erosion or loss of organic matter during high
rainfall events. (USDA Climate Hubs, 2013.)

Crop Yields and Climate Variability

Higher atmospheric carbon dioxide levels may reduce the quality
and nutritional value of crops such as alfalfa and soybeans, and
may also reduce grain quality. This is a health concern for
humans, as the nutritional value of food crops could be reduced.

Floods and droughts can damage crops and reduce yields. In
areas where soils dry out when temperatures are high, increased
irrigation will be needed. This may depend on whether sufficient
water is available, as drought could reduce water supplies, especially during times when irrigation is crucial to ensure good
yields.

Weeds, fungi, and pests thrive in warmer temperatures, which
may affect optimal yields. Mitigation measures could raise
production costs. Weeds, pests, and diseases may move into new
areas due to climate change, which may be detrimental to farms
and crops that have not been previously exposed to them.

Climate Change Impacts on Livestock

Livestock production and sales take place throughout the entire US. In 2022, the ten largest beef producers were: Texas, Nebraska, Kansas, California, Oklahoma, Missouri, Iowa, South Dakota, Wisconsin, and Colorado. 2023's top ten pork production states were: Iowa, Minnesota, North Carolina, Illinois, Indiana, Nebraska, Missouri, Ohio, Oklahoma, and South Dakota; and Georgia, Alabama, Arkansas, and North Carolina were the highest poultry producers (Cook, 2023; Statista, 2023; World Population Review, 2023).

Agriculture is also a significant contributor to developing economies, accounting for around 40% of agricultural GDP (Sejian et al., 2016). However, extreme weather is detrimental to livestock raising, as it reduces the quality and quantity of available forage.

Heat stress is one of the most challenging climate change effects on livestock, as this reduces milk and meat production, reproductive efficiency, and animal health, leading to increased costs for farmers. Animals lose weight and condition, and the quality of their products declines. Cows take longer to come into estrus and conception rates could reduce by 20%–27% in summer. Fetuses grow slower and miscarriages are more likely.

Livestock producers are affected by climate change impacts indirectly by:

- reducing the availability of animal feed and water resources
- increasing water demand for the cultivation of forage crops

- reducing the amount of forage, as well as reliability of supply and quality
- changing rangeland vegetation growth and types (Sejian et al., 2016).

One of the biggest issues faced by livestock farmers in the future will be how climate change affects both natural and cultivated forage production. Grassland ecology is already being affected by warmer temperatures, higher carbon dioxide levels in the air, and widely fluctuating water availability. Rainfall patterns and distribution ultimately determine forage quality in grasslands and this could become a major stumbling block for rangeland farmers. Lack of adequate water leads to animal weight loss, reduced disease resistance, and low reproductive rates.

Temperature and rainfall variability can also affect disease outbreaks, especially during warm, wet weather. Vectors that transmit diseases, such as ticks, mosquitoes, and biting flies, could survive year-round if winters become warmer. There are indications that such insects are changing their distribution ranges and spreading livestock diseases further. Parasitic diseases may also alter their ranges of occurrence. Foot-and-mouth disease (FMD) or bird flu outbreaks may affect large numbers of animals and further contribute to livelihood impacts for stock owners.

Very little has been done to address the vulnerability of livestock farming to climate change, although new technologies and techniques using artificial intelligence could help livestock producers with climate change adaptation.

Water Supply and Overabundance

Water covers over 70% of the Earth's surface. Life on Earth began in water and all living things need water. Climate change could affect freshwater availability, putting strain on areas already experiencing water stress. Other places may face more frequent flooding or increased risks of sea level rise and storm surges. Agriculture uses significant amounts of water.

Water is both a local and global resource. While regional changes in the quality and quantity of available water are felt at the local level, water is also a common, shared good. Water moves across countries and many water bodies are connected. This means that a local water problem can contribute to a much larger challenge. A global issue, like warmer ocean temperatures, can affect all of Earth's inhabitants.

Groundwater Depletion

Groundwater is an essential resource across the US and elsewhere. Where surface waters are scarce, people turn to groundwater to supply their hydration and agricultural needs. In the US about half the population, together with most people living in rural areas, rely on groundwater for drinking. Around 50 billion gallons of groundwater are used per day for agriculture (Water Science School, 2018).

When water is pumped out of the ground faster than nature can replenish it, this leads to groundwater depletion and water levels decline. This is a major issue across vast areas of the US The volume of stored groundwater is decreasing in many parts of the country. Beyond the obvious issue of dry wells, indications of groundwater depletion include:

- water quality deterioration
- increased pumping costs
- reduced water levels in streams and lakes

- land subsidence and sinkholes in areas of heavy withdrawal (Water Science School, 2018).

Excessive groundwater abstraction can lower underground water tables, which marks the point below which the ground is saturated. To abstract groundwater, pumps need to bring water up from below the water table. If the groundwater falls below the level of the pump, the well might need to be deepened. The water yield quantity and quality from the well will also decline.

It seems counterintuitive that groundwater depletion would lead to lower water levels in lakes and streams. The water in streams comes from groundwater seeping into its channel. Pumping groundwater can change the flow of water between an aquifer and a surface stream or water body. If too much water is pumped out of the aquifer, then the groundwater level may reduce to the point where it is too deep to adequately recharge the stream. This means riverine vegetation and wildlife habitat may be lost.

When water is pumped out of the soil, especially if this happens rapidly, it leaves a hollow into which the remaining soil collapses. This can either create sinkholes or the land itself might settle at a lower level. Removal of subsurface water often causes subsidence. The entire state of Florida, for instance, is underlain with carbonate rocks, which means that sinkholes could appear virtually anywhere. Sinkhole formation is accelerating—some are so large that homes, buildings, swimming pools, and roads disappear into their maws. Pumping groundwater for irrigation or urban water supply can produce new sinkholes in vulnerable areas.

One of the primary threats to groundwater quality is contamination with saltwater. Much low-level groundwater, together with that below the level of the ocean, is saline. There is more saline groundwater than fresh groundwater. Excessive pumping can move saline water found in the deeper levels of the soil into fresh groundwater closer to the surface diminishing the quality of the water.

Flash Floods and Urban Planning

You might have heard warnings for flash flooding being issued during high rainfall events. Flash flooding is usually caused by extremely heavy rainfall during a thunderstorm. Flooding usually begins within 3–6 hours of a heavy rain. Dam or levee wall breaks or mudslides can also cause flash flooding (National Weather Service, 2019). Over 1,800 people died due to flooding caused by Hurricane Katrina in 2005. In Orleans Parish, many people died as a result of water breaching broken levees (Wikipedia Contributors, 2023e).

These floods happen relatively quickly, and properties, roads, and footpaths can rapidly be overwhelmed by torrents of muddy water. Flash floods are more likely to occur in urban areas where there are no flood plains or natural vegetation to spread or hold back the water. Flooding can happen anywhere—in streams or creeks, city streets, or highway underpasses.

Melting Glaciers

When land-based glaciers melt, they cause sudden and deadly floods. Melting glaciers form depressions in the ground that fill with water. These are known as glacial lakes. Over the last 30 years, as climate change has increased the rate at which glaciers are melting, numerous glacial lakes have formed. Like concrete or earth dams, these lakes can break, releasing torrents of frigid water, rocks, and debris into the countryside downstream. This happens rapidly when a glacial lake is compromised.

More people have moved into the areas downstream of these lakes, especially in the Asian mountains of Nepal, Pakistan, and Kazakhstan. As a result, there are now around 15 million people at risk of experiencing one of these floods (Kuta, 2023). In Asia, there's little chance of getting an early warning. While there are also glacial lakes in the US Pacific Northwest and Greenland, these are either located far from human populations or where people are less vulnerable.

The impacts of devastating floods can be reduced by investing in early warning systems, running evacuation drills, limiting development in flood-prone areas, and halting global warming.

When glaciers melt into the ocean, the sea simply absorbs the icebergs that break off. On November 24, 2023, a 1,500-square-mile iceberg (roughly the size of the state of Rhode Island) called A23a became un-grounded after being stuck for nearly four decades since breaking off the icy Antarctic Filchner Ice Shelf in 1986. It is expected to move into the Southern Atlantic Ocean and continue to drift north toward Africa. Rising sea and air temperatures caused by climate change is contributing to a greater amount of ice loss and is known to contribute to the calving of ice that forms icebergs (Phillips, 2023).

Key Chapter Takeaways

- Weather and climate are not the same thing. Weather refers to short-term, daily forecasts about expected weather conditions, while climate considers long-term averages over time in a specific place.
- Certain weather variations can make one summer cooler than the next, for example, but the general trend is toward warmer weather as a result of climate change

over an extended period of time. This can be seen in long-term data sets.

- Climate is influenced by atmospheric and oceanic conditions, which are changing due to climate change and increased greenhouse gas emissions.
- Since 1980, the world has seen record-breaking summer temperatures leading to life-threatening heatwaves, unseasonal cold accompanied by heavy snowfalls, and changing marine ecosystems.
- Agriculture is experiencing longer and warmer growing seasons, which could alter places where crops are grown, reduce nutrient levels in food crops and livestock feed, and create conditions where weeds, diseases, and pests flourish and colonize new areas.
- Livestock production plummets when animals suffer from heat stress, and the quality of milk and other animal products deteriorates.
- Reduced water supplies due to fluctuating rainfall patterns or floods resulting from extreme weather events pose tremendous threats to livestock farmers in some regions.
- Freshwater scarcity is a growing problem. More people are depending on groundwater for drinking water and agriculture, and abstraction rates have become unsustainable, leading to negative impacts including dry wells, reduced water supplies, lower stream and lake levels, ground subsidence, and saltwater intrusion into aquifers.
- Flash flooding caused by sudden downpours during thunderstorms threatens urban areas as the water runs off hard surfaces, quickly engulfing buildings, streets, and vehicles. Melting glaciers create glacial lakes that

can burst, releasing vast quantities of frigid water, and flooding the countryside downstream.

If we continue to use and burn fossil fuels to power our lifestyles, we are contributing to greenhouse gas emissions in the atmosphere. By developing a greater understanding of how weather patterns everywhere are shifting, you will be better able to better appreciate the links between human actions and global warming. While there are natural causes for weather anomalies, our human influence is significantly adding to the mix.

As we finish considering the influence of climate change on natural weather patterns, it's time to examine our human footprint on the planet and discern fact from fiction in the blame game of climate change.

Human Contributions and the Blame Game

Nature has its own cycles and rhythms but the relentless pace of human-caused climate change is unparalleled.

As skyscrapers replace forests and carbon emissions cloud the skies, we must confront a particularly pressing question: Are we the architects of our own destruction? In this chapter, you will discover the truth behind humanity's contribution to the dire state of our planet.

The Role of Fossil Fuels

As mentioned previously, even in the early days of identifying global warming, 75 years ago, it was apparent that fossil fuel burning was the main culprit. Despite increasing evidence that humanity needs to reduce greenhouse gas emissions and find other sources of energy, the status quo continues largely uninterrupted.

Coal, Oil, and Natural Gas—Climate Change Impacts

Fossil fuels were created when prehistoric, carbon-based organisms, including trees and plants, died millions of years ago. Over time, their remains formed the carbon-rich deposits that are the coal, oil, and natural gas we use today. These are extracted and burned to provide the energy that powers the world's industries, economies, and modern lifestyles. Despite the advent of renewable energy options, around 80% of global energy still comes from fossil fuels (Client Earth, 2022). They are also used to make steel, plastic, and other products.

When fossil fuels are burned, they release the carbon stored within them into the atmosphere. Some, like natural gas, also release methane. In 2018, the IPCC found that 89% of all carbon dioxide emissions came from fossil fuel burning and industrial activities. These can be broken down as follows:

- Coal accounts for the greatest rise in global temperatures and is responsible for some 32.5 °F (0.3 °C) of the warming experienced on Earth so far. This makes coal the largest single cause of climate change.
- Oil releases a significant amount of carbon when it is burned and this accounts for one-third of all emissions. But that's not the end of its negative impacts. Ocean oil spills have severely impacted several marine ecosystems.
- Natural gas, often promoted as a cleaner energy source than coal or oil, is nevertheless a fossil fuel and is responsible for about one-fifth of global carbon emissions. (Client Earth, 2022.)

LNG, one of the world's biggest sources of energy, thought to be a replacement for coal, might be even worse for the environment than the energy sources it wants to replace and could be responsible for heating the planet faster than other known polluters.

New research from Cornell University shows liquefied natural gas, known as LNG, has an even bigger impact on climate change than burning coal.

"A broader conclusion is the need to move away from any use of LNG as a fuel as quickly as possible, and to immediately stop construction of any new LNG infrastructure," wrote Robert Howarth, the author of the analysis.

The findings have major implications for global climate goals and for the United States, which became the world's largest exporter of liquefied natural gas earlier this year (Cain C. 2023 Nov 30) (Howarth, R.W. 2023 Oct 24).

Transportation and Emissions

Transport creates 21% of global carbon dioxide emissions. Road transport is the biggest emitter, accounting for 74.5% of all transport emissions in 2018. Of that, most emissions came from passenger vehicles (45.1%) and freight trucks (29.4%). This means that road transport, as a whole, produced around 15% of total carbon emissions in 2018 (Ritchie, 2020). This is one of the reasons why many countries are promoting electric or hydrogen vehicles. Some countries are also improving their rail infrastructure, enabling more commuters to use trains, rather than using road transport options.

There have been endless discussions about the role of aviation in global carbon emissions. While not insignificant, it nevertheless contributed around 11.6% of all transport emissions, equivalent to 2.5% of total global emissions, in 2018. In case you're wondering, that equates to just under one billion tons of carbon dioxide a year. International container ships emit the same, accounting for about 10.6% of transport emissions (Ritchie, 2020).

Reducing transport emissions will be challenging. Car ownership is expected to escalate by 60% before 2070, while passenger and freight aviation could triple (Ritchie, 2020). Yet the world could easily shift to lower-carbon transport. The International Energy Agency forecasts that fossil fuel-powered vehicles could be phased out in China, the EU, and the US by 2040 (Ritchie, 2020).

The Fossil Fuel Lobby

It's hard to believe in these days of extreme weather, mass protests by young people concerned about the future, and the boom in renewable energy installations but fossil fuel producers are actively lobbying for business as usual. Some of the largest US corporations are fossil fuel producers. They make significant campaign contributions to political parties and their leaders in order to influence policy in their favor.

In the 2004 elections, oil and gas companies contributed a staggering $25 million to political campaigns; in 2006, they contributed $19 million, while electric utilities donated around $15–$20 million. Between 2003 and 2006, the energy lobby contributed over $58 million to political campaigns at the state level. Renewable energy interests were only able to contribute half a million dollars over the same period (Huang, 2022).

The energy lobby's most influential members are often some of the biggest polluters in the US. These companies continue to report enormous profits despite increasing emissions or releasing harmful chemicals at their refineries. Media reports indicate that

US government representatives have consulted with fossil fuel company executives on climate change and emission reduction issues, from signing the Kyoto Protocol to developing national energy policies.

The energy lobby has impeded open, honest discussions around energy issues at climate summits and conferences. While fossil fuel companies publicly ascribe to the Paris Agreement to keep global warming below 35 °F (2 °C), internal reports indicate that at least some of these companies have made contingency plans for warming of 37.4 °F (3 °C) by 2050 (Huang, 2022).

In 2019, British Petroleum (BP) spent millions on advertisements promoting its low-carbon energy and cleaner natural gas while spending 96% of its annual budget on fossil fuels. After complaints by environmental groups, BP withdrew its misleading advertising (Client Earth, 2022).

Impact of Deforestation

Deforestation is increasing globally. Forests are logged for timber and to expedite palm oil plantations, agriculture, and mining. Using fire to clear land further threatens forests. Globally, around one billion acres (420 million hectares) of natural forests were lost between 1990 and 2020, and a further 24.7 million acres (10 million hectares) disappear annually (The Royal Society, 2022). One of the hardest-hit areas is the Brazilian Amazon, which has lost 20% of its forest cover since the 1970s. Although deforestation has slowed dramatically in recent years due to new forest protection policies, between January and August 2023, 1,225 square miles (3,712 square kilometers) were felled (Butler, 2023). Scientists are concerned that persistent deforestation twinned with climate change may convert the rainforest into a drier, savanna-like ecosystem. This will negatively impact rainfall patterns, carbon storage, biodiversity, and communities.

The situation has become so serious that the United Nations has warned of imminent ecosystem collapse if agricultural expansion in the Amazon continues unchecked.

Biodiversity Loss

Forests are vast storehouses of biodiversity supporting thousands of species of plants and animals. These forests could support species of plants and animals that have yet to be identified. Deforestation has fragmented forests, negatively affecting species that need vast tracts of land for survival, and isolating plants and wildlife. Some unique creatures threatened by Amazonian deforestation, for example, include orchids, bromeliads, butterflies, bees, ants, turtles, and jaguars. Forest loss also changes the species matrix, as some will inevitably be lost. Generalist species better able to adapt to disturbed habitats, such as foxes, deer, and storks, are increasingly colonizing fragmented forest patches.

Soil Degradation and Erosion

The top layer of soil—the topsoil—is the most fertile and enables the survival of trees and plants. It takes thousands of years to form just one inch of topsoil (Begum, 2021). When soil becomes degraded it is no longer able to support plants and animals. Beneficial soil life disappears when there are detrimental changes in the soil and it loses its fertility.

Soil not only enables plant growth, it anchors plants. Healthy soil purifies water by filtering out pollutants, regulates water flow, and helps to prevent flooding. It supports organisms that kill harmful bacteria. It is an ally in the fight against climate change, as it is the Earth's second-largest carbon store after the ocean.

While soil degradation happens naturally, it can be accelerated by human actions. Soils have been compromised by deforestation, intensive cultivation, overgrazing, forest fires, and construction. When forests are felled to make way for farmland, this reduces certain nutrients present in the soil, as these are found mainly in decaying organic material beneath the trees. Land conversion of this kind reduces the carbon storage capacity of the soil by as much as 50%–75% (Begum, 2021).

Soil degradation can lead to erosion, landslides, and floods, as well as desertification, pollution, and reduction of the global food supply. People in low-income countries may also need to migrate to safety from catastrophic events and to find more fertile land.

The Amazon Rainforest—The Planet's Threatened Green Lungs

Slash-and-burn agriculture is threatening one of the largest rainforests on Earth—the Amazon. This is a very destructive method of land clearing and has already devastated around 20% of the forest (Tree Plantation, 2023). This has profound implications for climate change. Thousands of trees in the Amazon absorb carbon dioxide from the atmosphere and are accordingly one of the Earth's biggest carbon storehouses. They are often referred to as the planet's "lungs." This not only reduces the effects of climate change; it also determines rainfall patterns across the Americas.

There are several tree species threatened by slash-and-burn agriculture in the Amazon, including the Brazil nut, mahogany, rubber tree, rosewood, and kapok.

Industrialization: A Two-Edged Sword?

Historically, productivity has been allied to rising emissions. An event that appeared to confirm this was the reduction of emissions following the 2008 global financial crash (Serin, 2022). The link between economic activity and climate change occurs because coal has traditionally been used to provide energy for production. This is a primary climate change driver, making it imperative to transition to other energy sources, which could reduce climate change impacts while maintaining robust economies.

There is some debate about whether economic growth will be possible under increased climate change. Unmitigated climate change may reduce economic development but stopping economic growth entirely could also make things worse. Recessions tend to impede more environmentally friendly production methods, for instance. Traditional ways of measuring economic growth—such as GDP, which measures all the goods and services produced by a particular economy in a year—do not take environmental benefits and costs into account.

Economic growth and development need not increase carbon emissions. Technological improvements could reduce the energy inputs and materials required for production. Some high-income countries such as Germany, the UK, and the US are moving away from coal to low-carbon energy sources, improving climate change policies, and shifting their economies from manufacturing to service-based industries.

The cost of low-carbon energy relative to that generated by fossil fuels will determine the extent to which economic growth can be decoupled from increased emissions. There have in recent years been significant reductions in the cost of renewable energy, which could help to facilitate this transition.

Investing and increasing the labor force can also improve economic growth. Renewable energy provides three times more jobs than fossil-fueled energy development. The increased labor requirements of renewable energy could provide more jobs, thereby growing the economy.

Sustainability can drive a net-zero economy while delivering continued economic growth. This requires changes on both the supply and demand side of consumption, as well as individuals moving toward more sustainable lifestyles. The effect of emissions embedded in imported goods and services also needs to be considered. Sustainable development means that economies can grow without raising emissions. Economic benefits include avoiding climate-changing emissions altogether, as well as reducing the costs of climate change mitigation. Consumption of non-material goods, such as art, literature, music, and psychological insights creates value and promotes well-being without natural limits. Expanding the intellectual economy can therefore fuel economic growth.

Pollution and Public Health

Many urbanites are forced to breathe air well below the World Health Organization Air Quality Guidelines (WHO AQG) (Kelly & Fussell, 2015). Modern pollutants, unlike the infamous pea-souper fogs of Victorian London, are largely undetectable, yet have chronic health effects. Modern pollution contains both nitrogen oxides and particulate matter. Particulate pollution is generated by fossil fuel burning and is derived from road traffic, ships, smelters, factories, desert dust storms, and coal-fired power stations. Burning wood for fuel in developing countries is a major source of particulate pollution.

Several studies worldwide have established that particulate pollution is increasing mortality from respiratory and cardiovascular conditions, especially where the deceased were exposed to heavy pollution loads for longer periods of time. In the Western Pacific and Southeast Asia, these deaths were linked to heavy industry and pollution hotspots. In Europe, air pollution consistently causes premature deaths in middle- and high-income countries. Between 64% and 92% of Europeans exposed to particulate pollution above the WHO AQG were found to experience reduced life expectancy. More recent research indicated that particulate pollution is a contributing factor to atherosclerosis. Long-term exposure to traffic-related particulate pollution was found to increase the incidence of asthma and reduce lung function in young children (Kelly & Fussell, 2015).

Particulate air pollution is now believed to cause a wider range of health difficulties, affecting diabetes, neurodevelopment, and cognitive function. Older people living close to busy roads have a higher risk of developing Alzheimer's Disease (Kelly & Fussell, 2015).

In Swiss and US studies, it was found that improvements in air quality significantly reduced these health problems and risks. Improvement was evident almost as soon as exposure to particulate matter was reduced.

Innovations for Cleaner Industries

Economic development and competitiveness generally shape technological and other innovations. Few individuals are prepared to incur the costs of reducing the impacts of pollution, even if doing so would be of benefit. To prevent global temperatures from rising above 35 °F (2 °C), global greenhouse gas emissions would need to be reduced by between 50% and 80% from pre-1990 levels (Rubin, 2018).

Actions required include:

- reducing energy demand across all economic sectors
- improving efficiency so fewer fossil fuels are required
- replacing fossil fuels with low-carbon alternatives like biomass, wind, and solar
- capturing and storing carbon dioxide emitted to prevent its release into the atmosphere (Rubin, 2018).

Technology can help to promote mitigation methods such as reducing land-use change and deforestation or avoiding emissions from nitrous oxide emitted during agricultural activities. However, the degree of change required will take decades because technological progress is complex and is often dependent on technological advancements in other areas. New technologies work better if there is widespread adoption, which all takes time. Many of the required technologies are also either too costly or don't yet exist.

China has become the world's largest carbon emitter but is investing heavily in renewable energy technologies, including solar and wind generation, together with nuclear power. In the process, it has become the largest manufacturer of solar PV cells. The country is working to develop clean coal, and carbon capture and storage systems. By contrast, the US has reduced government spending on energy R&D over the last 30 years, while spending on renewable energy R&D is lower than that for other economic sectors. It is also much lower than that for other countries when considered as a percentage of GDP. The private sector will need to fill the gap (Rubin, 2018).

However, there are simpler methods that can be used while we wait for technology to catch up. Changes to urban planning and development could reduce future energy demands for transport, for example, as well as in buildings and homes. Incentives could be created by utilities to encourage consumers to use less energy.

One of the difficulties is that there are few current markets for efficient, low-carbon technologies that would slow climate change. Costly actions by companies and individuals that provide little tangible benefit to them are unlikely to gain traction. Governments will need to create or improve markets for emission-reducing technologies. Policies can take the form of mandatory requirements or voluntary measures while limiting certain activities.

Overcoming Denial

According to Naomi Klein, writing in *This Changes Everything*, many climate deniers take the stance that, if it's real, the political fallout from climate change would be severe. These groups also fear the economic and social implications of climate change and are working to disprove the science (Klein, N. 2015 Aug 4).

Debunking the Myths

Here are some facts about climate change you can use when faced with those who deny its existence:

- Some argue that there is no global scientific consensus on climate change. However over 97% of the scientific community affirms that this is not only a reality, it is caused by human actions. The IPCC includes 2,000 independent scientists.

- Contrary to denialist belief, volcanoes do not emit more carbon dioxide than humanity. They actually emit less than 1% of total carbon dioxide emissions, so humans are by far the greater emitter.

- While it is true that the Earth has gone through warmer and colder periods, the current increase of 33.9 °F (1.1 °C) since pre-industrial times is happening faster than increases created by the natural cycles of the planet.

- Solar cycles change and there are times when the Earth is warmer. For the last 35 years, the effects of the sun on our planet have been greatly reduced, yet global warming is increasing.

- It is not true that the effects of climate change will take so long to be noticed that your children won't see them. Since 1980, there has been a marked and very obvious rise in global temperatures. Every summer since 2014 has broken the previous one's temperature record. We are also seeing sea levels rise, oceans acidifying, the likely imminent disruption of the AMOC and intensifying natural disasters like hurricanes, tornados, fires, and floods, which are all fueled by global warming. Climate migration and the disappearance of fauna and flora are also occurring.

- A temperature change of a few degrees might not seem like much but natural systems, wildlife, and people who depend on a stable climate for survival are adversely affected. One example is birds whose eggs hatch at the same time as caterpillars so that there is a food source for their young. Because of climate change, the caterpillars are emerging earlier. By the time the chicks hatch, there are fewer caterpillars for them to eat. Some bird populations are declining precipitously because of this situation. (Acciona, n.d.)

The Media's Role in Climate Change Perceptions

The media thrives on sensationalism—and the local and regional impacts of climate change have largely become yesterday's news. Climate change coverage focuses on extreme weather events and there is little opportunity to delve into the issues around the causes of global warming. However, factual reporting can inform people about climate change: A recent US study found that this persuaded even global warming skeptics to support adaptation policies (Russell, 2022).

When the media reports climate change, it uses emotive language that provides little balanced information and creates fear or apathy. In 2022, Just Stop Oil vandalized artworks and soon everyone knew everything there was to know about the organization (Russell, 2022). Focusing too much on the negative impacts of climate change also creates climate anxiety and apprehension. It's important to focus on the positive steps and viable solutions that have been taken and implemented.

Peaceful climate protests, such as the ones in which hundreds of young people around the world now participate, are a good way to highlight the issues. It's important to focus on climate justice and inspire others.

The Importance of Science Education

Climate change is such a pressing issue that educating young people about global warming is essential. Today's children are highly likely to experience the full force of climate change effects and need to be prepared.

Over the last 20 or 30 years, scientists have made predictions about the impact of climate change but it is people on the ground who are actually experiencing the consequences. There is little public concern about the difficulties and suffering climate change may cause. Surveys in France in 2017 and 2018 indicated that adults were more concerned about the potential impact of climate change on employment, although 57%–59% of high school students and graduates expressed an interest in working in environmental fields (Pontifical Academy of Sciences, n.d.). Besides learning about climate change, education facilitates behavioral changes that enable climate change adaptation and prevent warming.

In 2018, France launched the Office for Climate Education, which develops resources and tools based on IPCC reports for teachers. The organization works with partners in several countries. Initiatives in California, France, and India are producing climate educational tools for students and educators. Teacher training is being done in Benin, Chile, and Tunisia.

Your Daily Environmental Challenge

When faced with something as enormous as climate change, it can be difficult to visualize what you could do as an individual. However, you can easily reduce your planetary and environmental footprint, together with the emissions you and your household generate. Create a Daily Change Challenge for your family. List seven things you can do once a week to make a difference to the environment and reduce climate change. Examples include "Meatless Mondays" or "Walk-to-Work Wednesdays."

Here are a few things you could add to your list:

- Switch to refillable bottles rather than single-use plastic ones.
- Take a reusable porcelain mug to your local coffee shop to save Earth's natural resources and reduce trash.
- Take a shorter shower to reduce water consumption.
- Replace plastic straws. Invest in stainless steel or bamboo ones you can carry with you.
- Choose a reusable grocery bag wherever possible instead of the plastic ones that not only end up in landfills but require petrochemicals to produce.
- Buy local goods and produce to reduce transport emissions created by moving goods over long distances.
- Find out more about how to recycle properly and implement what you learn to ensure that reusable goods are recycled. Wash your recyclables to prevent contamination. This saves planetary resources.
- Hang laundry in the sun to dry, as this reduces energy usage.
- Donate old clothing rather than throwing it away.
- Buy in bulk and use mason jars and reusable containers to reduce packaging.
- Purchase items that are ethically and sustainably produced to save resources and support companies with a better environmental ethic.
- Buy secondhand home décor, dishware, clothing, etc.
- Take a cold shower and turn off the lights when you leave a room, as these save energy usage.

- Use eco-friendly transportation by walking, biking, or carpooling.
- Sign an environmental petition.
- Reduce food waste and don't buy more than you will eat.
- Make your own compost to reuse kitchen scraps, garden clippings, and other organic waste as a soil amendment.
- Switch to paperless communication as much as possible.
- Buy a tree and plant it in your yard to soak up carbon dioxide. (Cooley, 2021).

Key Chapter Takeaways

- Fossil fuels are high-carbon deposits created prehistorically. They release the stored carbon when they are burned. Coal emits the most carbon, followed by oil and natural gas. Fossil fuel producers contribute enormous sums to political campaigns to protect and further their financial interests.
- Rainforests are the most biodiverse ecosystems on Earth but deforestation is destroying this tremendous wealth of plants, birds, and wildlife. These ecosystems sequester large amounts of carbon and influence rainfall patterns worldwide. Soil may also hold significant amounts of carbon but this is reduced as it becomes degraded. Degraded soils are also less fertile and prone to erosion.
- Rapid economic growth can lead to increased carbon emissions but this is not always the case. Transitioning to a low-carbon economy can stimulate economic growth and provide jobs in fields such as renewable energy and developing technologies to reduce emissions.

- Particulate pollution from fossil fuel burning has many negative impacts on human health, causing respiratory and cardiovascular conditions. Sources include coal-fired power stations, vehicle traffic, shipping, smelters, factories, desert sandstorms, and burning fuelwood.
- When faced with climate change deniers, several points can be made to debunk common myths. Climate change is real and is having negative impacts on people across the world. The media often focuses on sensationalism and rarely reports on the science behind climate change, choosing to focus on extreme weather events or the shocking actions undertaken by some activists. Climate education is essential to prepare young people for the future under climate change. Teachers and educators are being provided with training and materials to facilitate this.

Climate change is real. Scientists and natural systems are confirming that the planet is warming at an unprecedented rate. A large portion of the Earth's population have personally endured climate change impacts, whether they had to evacuate because a hurricane was barreling toward their coastal region, were hospitalized during a heatwave, had their home destroyed in a tornado, or lost their livelihoods when their business premises flooded.

Ordinary people are being traumatized by climate change, left to sift through their belongings and argue with insurance companies in the wake of a devastating wildfire or flood. Farmers watch their livelihoods burning away during a drought. The rains, when they come, are so torrential that they sweep away what's left. These are the "silent sufferers" no one sees. No media event tracks them shoveling mud out of their houses or gazing despairingly at acres of fire-blackened land that once supported large herds of cattle. No one watches them pack their belongings and leave the devastation in search of a better, less disruptive, life. In Chapter 5, you will meet more of these people, climate change's silent casualties, whose lives will never be the same.

Climate Change Reality Check

"We do not inherit the Earth from our ancestors; we borrow it from our children"

— Native American Proverb

Hello Environmentalists

Our mission is to make Climate Change and its impacts understandable by everyone. Everything we do stems from that mission. And, the only way to accomplish that mission is by reaching… well… everyone.

That is where you come in. Most people do, in fact, judge a book by its cover (and its reviews). So here's my ask on behalf of a struggling Environmentalist To-Be you've never met.

Please help that new Environmentalist by leaving this book a review.

Your gift costs no money and less than 60 seconds to make real, but can change a fellow Earthlings life forever.

Simply scan the QR code to leave your review.

Your review could help… This book isn't just another read on

environmental issues – it's a deep dive into the critical challenges our planet is facing due to climate change and a detailed look at how global warming is changing our weather, leading to costly disasters, and what we can do about it.

We value your opinion and we're hoping you could share your honest review. Your insights are key in helping others understand these complex issues and what they can do to help.

Thank you for helping to spread awareness and inspire change.

The Silent Sufferers—Ordinary People in the Climate Crisis

Some 25 million people are already being displaced by climate events each year. Between 2030 and 2050, it is estimated that some 250,000 people will die from climate change effects. The most vulnerable are poor people, 75% of whom harvest natural resources and undertake subsistence agriculture for survival (Mercy Corps, 2021). However, climate change is altering rainfall patterns, deepening droughts, and intensifying severe weather. Climate change is a threat multiplier, which increases soil erosion, the incidence of diseases like malaria, malnutrition, and the effects of poor air quality.

The world's poorest have no insurance, few reserves, and often cannot recover from these events. Communities fracture and this may cause conflict over scarce resources, migration, and dependence on humanitarian aid.

In this chapter, we will unpack the human stories, discuss the economic repercussions, and detail the opportunities resulting from the changing dynamics of our planet's climate.

Stories From Ground Zero

In 2022, 70% of communities in the US experienced devastating heatwaves, droughts, floods, wildfires, hurricanes, and tornados (The Wilderness Society, 2022). When this happens to you, climate change becomes personal. Climate change doesn't respect human-imposed borders and often affects vulnerable, underserved communities the most.

One in six Americans live in areas with significant wildfire risk. If you live in any area that has experienced severe wildfires in recent years, you're probably living in a constant state of hyper-awareness, watching for wisps of smoke on the surrounding hills and the smell of it in the air. Numerous communities have lost nearly everything they own, or been forced to flee their homes and businesses when wildfires roar across the countryside, devouring everything in their path. The state of California is at highest risk because of its large size and mild climate.

Paradise, California, was once so thick with trees that they obscured the town from view. All that changed after 2018's devastating Camp Fire, which killed 85 people and destroyed 18,000 buildings in the first four hours. Since the fire, the population of the town has reduced from 26,000 to 10,000 inhabitants (Wikipedia Contributors, 2023c). On August 8, 2023, on the Hawaiian island of Maui, wildfires near the town of Lahaina incinerated everything in their path and killed 99 people (Childs, 2023). Those who escape the infernos glance anxiously over their shoulders, wondering if they'll be next.

Extreme heat is threatening shady avenues and parklands across the city of Vancouver, Washington, as temperatures spike to unprecedented highs. During the summer of 2021, city trees like big leaf maples and iconic Douglas firs felt the heat. While some recovered, many are dying, changing the character of the city's neighborhoods.

Extreme heat, droughts, and flooding have become the norm in border towns like Nogales, Mexico. When heavy rain follows dry spells, flash floods abruptly turn streets into fast-flowing, muddy torrents. Residents hunker down until the worst has passed, praying that their homes and loved ones will be safe.

Toward the end of September 2023, the remnants of Tropical Storm Ophelia surging across the Atlantic Ocean joined forces with a mid-latitude system arriving from the west. The resulting storm stalled over New York City for 12 hours, unleashing between six and eight inches of rain over different parts of the city. Flood waters surged across the metro, flooding subways and railways, and leading to rail transport disruptions. The city's three airports also experienced disruptions, with La Guardia having to close temporarily after its Marine Air Terminal flooded (NOAA, 2023b).

Survivors of Hurricane Fred, which hit Durham, North Carolina, in August 2021, recall the devastation wreaked by the storm that flooded homes and businesses, killing six people in their quiet town. A resident who hails from India says her hometown regularly experiences floods and she worries about the effects of climate change on both Chennai, India, and Durham. A climate change advocate, she believes that severe weather will continue if global warming remains unchecked.

Social-Cultural Effects of Disasters

Natural disasters take us back to basics. We're forced to change our worldview and examine our values and priorities. Victims may feel grief, anger, or despair as they lose friends and relatives, see their homes in ruins, or watch their livelihoods vanish overnight. Communities who have previously experienced such events may be more resilient but, for those whose lives are disrupted unexpectedly, it can come as a tremendous shock. If the emotions triggered by the event linger, the survivors may experience PTSD. Disasters happen to everyone across all cultures and ideologies. Compassion for others is a natural human response, motivating people to assist. Donations to humanitarian organizations can also help to relieve the effects of disasters.

Resilience and Community Bonding

The climate crisis is likely to usher in a perfect storm. The effects we're seeing are just the tip of the proverbial iceberg that will likely change ordinary lives forever. Billions of people will lose their homes, livelihoods, and communities. We're likely to see more famines, wars, and poverty. One of the ways we can prepare for what is probably inevitable, given the general apathy, is to build more resilient communities.

Resilient communities adapt swiftly to rapidly changing circumstances and are economically stable. They are more likely to be able to sustain themselves in the event of a disaster.

Here are some of the ways resilient communities are equipping themselves for a less stable future:

- Building more resilient buildings with:

  - deep foundations capable of withstanding earthquakes and high winds
  - renewable energy to supply power when utilities and grids are down

- ○ flexible materials that can buckle without breaking
     during extreme weather events
- ○ better and natural thermal efficiency to keep
     buildings warm in winter and cool in summer.

- Building more durable infrastructure with sustainable
  materials using new technologies like bendable concrete
  that is shock-resistant, or recycled materials. More
  robust infrastructure helps communities recover after
  crises.
- It's important to include environmental education in
  school curriculums to encourage innovative responses to
  climate change and disasters. Young people from smaller
  communities will then be more likely to return home
  after receiving university education elsewhere.
- Healthcare systems and infrastructure can become
  strained by extreme weather events. Communities with
  coordinated responses can ensure that affected
  individuals and households receive timely assistance.
  (Fletcher, 2023).

The Rise of Climate Refugees

Since 2008, an average of 26.4 million people a year have been displaced by floods, droughts, earthquakes, or windstorms (Prange, 2022). While some displaced people find refuge in their own countries, others need to move further away.

Most countries have no protocols for coping with this influx of climate migrants, and their response has been limited and often inadequate. Both Norway and the UN have identified natural disasters as a significant cause of migration. It has been suggested that climate action should feature in discussions concerning people who are displaced due to extreme weather, rising seas, or desertification.

Climate migrants do not have formal refugee status and are therefore not protected by international laws governing the treatment and status of those fleeing war, famine, or persecution in their home countries. Although terms like "climate refugee" or "environmental refugee" have been used for decades, they have yet to be legally defined, which means people fleeing natural disasters or islands that are gradually sinking beneath the waves have no guaranteed protection. The main difficulty appears to lie in distinguishing climate migration from migration due to other causes related to economically depressed economies or rural-urban movement.

On the flip side, a particularly adverse consequence of climate change may mean that some people are unable to move, becoming what some call "trapped populations" (Schewel, 2023). This is when a lack of resources dooms people to remain in place, despite the potentially severe impacts of global warming. In some cases, people will resist the attempts of governments to move them away from problematic areas.

Legal responses to climate migration may need to become more nuanced, as there are several diverse ways in which climate change could impact a particular community or population. This could, for example, allow the export of labor so that the children of potentially trapped families could migrate to a safer place with better economic opportunities.

Migration Patterns and Challenges

Regions that are the most vulnerable to climate change include sub-Saharan Africa, southern Asia, and Latin America. About three-quarters of the population of the developing world reside in these regions and the World Bank anticipates that some 143 million people there could be displaced internally by 2050 (Prange, 2022). Those close to an international border may move to another country.

People already displaced internally for reasons other than a climate-related disaster often live in places prone to severe weather events. They might need to move a second time, which reduces their chances of returning to their places of origin.

There may be situations where climate change impacts intersect with armed conflict and violence. In the Sahel, a region just south of the Sahara, communities and individuals were already being displaced as climate change and poor conservation practices were leading to desertification. In recent years, some three million people have been uprooted due to worsening indiscriminate violence. This is creating a severe humanitarian crisis (UNHCR, 2022).

International Response and Responsibility

Across the world, climate change disproportionately affects indigenous people, people of color, and women—around 80% of climate change refugees are women. As caregivers, they are less likely to evade poverty, so the chances of recovering from climate change impacts are greatly reduced (Greenfield, 2022).

By refusing or delaying efforts to cut emissions, wealthy nations have fueled the climate crisis, which is creating refugees endeavoring to escape its consequences. A 2020 report by Oxfam International established that the richest 10% of the global population accounted for 52% of greenhouse gas emissions between 1990 and 2015. Annual emissions also grew by some 60% during that time, with the richest 5% of the world's population causing 37% of this growth. The increase in emissions by the richest 1% of the world's population was therefore three times more than that of half of the poorest people on Earth (Ratcliff, 2020).

World leaders only began to consider climate migration in 2015, at COP21, held in Paris. Although groups have been formed to address climate migration, little progress has been made. The main reason for inaction has been a lack of funding for climate adaptation as a driver of migration.

The US has, though, set up a task force to consider climate migration, including the legal status of these migrants. However, the US is also experiencing climate displacement within its own borders. Following Hurricane Katrina in 2005, for example, over a million people fled the Gulf Coast and were slow to return, especially if they were people of color. Indigenous groups across the US are struggling with extreme heat and less rainfall on their lands but are largely obliged to navigate complex relocation initiatives without assistance. One suggested solution is to create a framework for relocation that affected indigenous groups could adopt. By changing certain policies, the Federal Emergency Management Framework could also issue funding directly to affected communities to enable relocation efforts.

We need to start observing how the places where we live are changing and decide how we are going to adapt to these new situations. It's easy to assume that because you have means you will be left untouched, but climate change impacts can happen to anyone at any time.

Health Implications of Climate Change

Climate change means that certain disease-causing insects are moving into new territory. People are also becoming more stressed, anxious, and depressed as they either experience or are threatened by climate change impacts. Healthcare facilities are often inadequate for the vast numbers of people who require assistance after a severe weather event or natural disaster.

Vector-Borne Diseases and Warming Climates

Insects such as mosquitoes and ticks that spread disease are referred to as vectors, which can spread infectious agents or disease-causing bacteria between animals or from animals to humans. Warmer weather is enabling these vectors to colonize new areas. Higher rainfall creates more standing water in which insects breed. Even droughts improve conditions for vectors as pools of standing water occur when rivers and streams dry up and become a breeding ground for disease-carrying mosquitoes. Temperature changes might alter the behavior of some vectors. For example, higher temperatures modify the way mosquitoes bite, so mosquito nets are less effective.

The problem is expected to escalate—by 2050, disease-carrying mosquitoes will reach around 500 million more people a year than they do today. Mosquitoes alone spread diseases such as malaria, dengue fever, Zika, and West Nile virus. Studies have discovered that mosquitoes die when water temperatures exceed 93 °F (34 °C) and air temperatures are over 104 °F (40 °C). However, they are hardy survivors and have taken to hiding in cooler places like cement water tanks and household pitchers.

Ticks cause several diseases, including Lyme disease, and warmer winters have enabled their spread in many areas. Tropical ticks that cause encephalitis appear to be moving into subarctic regions in Asia and Europe. They were found in Germany in 2019 and appeared to have survived the winter (Dangour, 2022).

Minimizing the Risks

Besides mitigation, there are steps that can be taken to adapt to this new reality:

- minimizing exposure to disease-carrying insects by closing doors and windows, and wearing protective clothing
- providing healthcare access
- managing vector-borne diseases
- improve disease awareness and detecting outbreaks early
- managing wetlands and eliminating breeding sites close to human residential areas
- considering new control methods
- developing vaccines (Dangour, 2022)

Mental Health Impacts

Climate change is likely to raise the frequency of extreme weather, as well as slow-onset disasters like droughts. Since ancient times, it has been clear that weather changes affect humans physically, with some individuals being more sensitive than others. These people may respond physically, mentally, and emotionally to weather patterns.

Mental health effects can manifest quickly with incidents such as heat waves, or more slowly, as a result of prolonged droughts, deforestation, rising sea levels, or migration. Those affected often display mood disorders like anxiety, depression, and aggressive behavior, as well as substance abuse, increased suicidal tendencies, and PTSD. Vulnerable populations who are more likely to experience climate change impacts will often suffer from poor mental health as a result. Extreme weather might sever social ties and prevent community activities, making things worse.

Enduring extreme weather events can lead to increased symptoms of distress, raised suicide rates, and increased psychological disorders. As most of these events spark natural disasters, PTSD is often experienced. Those who are more inclined to respond to weather in general are likely to be more affected by gradual climate change impacts such as sea level rise, landscape changes, and the loss of environmental landmarks.

Eco-psychiatry could emerge as a new field of research and study as we work to understand the effects of climate change on human bodies and minds.

Strain on Healthcare Systems

Climate change is detrimental to our collective health, as evidenced by these statistics:

- Between 1997 and 2006, approximately 5,600 heat-related deaths occurred across the 297 counties where 60% of Americans reside.
- Hurricane Katrina resulted in nearly 2,000 deaths and displaced around 1.5 million residents in 2005.
- Flooding affected over 74 million people worldwide in 2016.
- Fine particulate material in the atmosphere led to 1.8 million deaths globally in 2019.
- In Washington and Oregon, the June 2021 heat wave resulted in 600 excess deaths in one week.

- Unusually hot or cold temperatures cause some five million deaths each year. (Seervai et al., 2022).

Climate change and its related injuries and diseases affect the entire healthcare system. It makes it difficult for people to access quality health services. Healthcare costs are increasing, partly due to the impact of climate change. A severe weather event may increase emergency department visits, hospital admissions, and regular doctor's visits. Wages may be lost due to ill health. In 2012, healthcare costs from ten climate events totaled $10 billion when all these factors were taken into account (Seervai et al., 2022).

Extreme weather events disrupt healthcare when hospitals are evacuated, facilities are damaged, power outages occur, and damaged roads or transit systems make it difficult for people to reach medical care. The closure or destruction of hospitals means others need to fill the gap, resulting in overcrowding. Disruptions to supply chains can cause shortages of essential medical equipment and medication.

Climate change makes the work of healthcare personnel more challenging, potentially leading to burnout. As more people fall ill or are injured due to climate change, more healthcare workers will be needed. As new healthcare challenges develop, healthcare workers will need to be trained in identifying and treating these risks. However, not all workers can cope with the difficulties of caring for people under climate change scenarios. An Australian study found that about a third of healthcare workers in underserved communities were moving to parts of the country that experienced fewer climate change risks (Seervai et al., 2022).

The US needs to build resilience by investing in resources to prepare for climate-related healthcare needs while reducing the sector's own contribution to climate change.

Economic Difficulties

Several economic sectors are experiencing climate change effects, especially tourism, which is a big employer. Although global markets are reflecting climate disruption, opportunities for green jobs are increasing.

Climate Change and Tourism

Tourism is being particularly hard hit by climate change impacts. The sector employs numerous rural people, either directly or indirectly, who in turn support their families and extended families. Some of tourism's climate impacts are direct, including a lack of snow at mountain resorts, the destruction of tourist attractions and beaches in hurricanes or as a result of sea level rise, or when tourist facilities burn down in a wildfire. Floods and landslides can affect tourism infrastructure, attractions, and accommodations.

Other tourism impacts are indirect but nevertheless significant. These include water shortages; damage to transport infrastructure such as roads, railroads, and airports during extreme weather events; and landscapes becoming denuded in the wake of heat waves, droughts, or floods. Biodiversity loss could impact tourism niches dependent on wildlife viewing, while climate change erodes low-lying atolls and bleaches coral, threatening island tourism. Climate change may shorten tourist seasons, increase prices due to rising maintenance and operational costs, and reduce natural beauty.

Economic growth may decline in certain areas, leading to a reduction in tourism as fewer people can afford to travel. Negative environmental impacts may destabilize countries, leading to conflicts over resources and land. This all has the potential to impact the tourism industry adversely.

Fluctuations in Global Markets

Climate change could disrupt or slow economic output and business opportunities across the globe. Frequent, intensely severe weather can damage factories, supply chains, and transport infrastructure. Water could become scarcer and more expensive, a situation that droughts could exacerbate. Raw materials could become more costly and harder to find. Climate variability makes it difficult for businesses to forecast prices for production, raw materials, energy, transport, and insurance. Some products might lose their markets or become obsolete, like a ski resort without snow, or restaurants curtailing their menus due to shortages of certain foods.

Carbon pricing or policies favoring competitors can affect a business's profitability. Companies might suffer reputational damage if they are not perceived to be environmentally friendly. Investors are increasingly concerned about stranded assets that fall out of favor to become undesirable, such as fossil fuels or a housing project built near an eroding shoreline. Flooding in one country could impact raw materials or manufactured components intended for another. Data centers could face increased cooling costs (Cho, 2019).

As fossil fuels like coal lose popularity, this could have knock-on effects for banks and investors. For instance, the US coal industry was valued at around $37 billion in 2011. Today, it's closer to $2 billion. General Electric overestimated demand for natural gas as opposed to renewable energy, costing its investors over $190 billion between 2015 and 2018 (Cho, 2019).

Protection

When making investment decisions, individuals and businesses should consider climate change implications. This means not purchasing coastal properties that could potentially be impacted by sea level rise, or buying land with high risk of wildfire. Fire and flood insurance availability and cost should always be considered, and investment portfolios should be diversified. Investors and banks are now including climate change in their risk assessments and factoring it into their portfolios.

Governments should prepare for disasters by shoring up infrastructure, moving people out of harm's way, and improving healthcare services and water resources.

Green Job Opportunities

To avoid the worst impacts of climate change, we need to transition to a low-carbon economy as quickly as possible. This will also deliver numerous green jobs in new economic sectors. The circular economy repairs, reuses, and recycles goods, which all create green jobs.

Green job sectors include:

- renewable energy
- agriculture, especially organically produced food and beverages
- eco-friendly design
- ecotourism
- electric vehicles and other low-emission forms of transport

Green careers include:

- solar panel installation technicians
- drone engineers
- operators of renewable energy power plants
- recycling plant technicians
- sustainability supervisors
- environmental scientists
- smart network managers

(Iberdrola, 2020.)

Watch and Empathize

Climate change can seem like a distant nightmare if you haven't yet encountered its realities firsthand. The following documentaries will help you to appreciate how climate change is affecting ordinary individuals and disrupting normal lives.

Bangladesh's Climate Refugees

Bangladesh is likely to suffer some of the most severe climate impacts in the world—and it's already happening. Within the next 30 years, as much as 20% of the country will disappear beneath rising seas and raging rivers (DW, 2019). The Meghna River has become just one of these raging torrents, which is now 5.5 miles (9 kilometers) wide in places. The river is tearing villages apart and ripping houses off their foundations, forcing the inhabitants to leave while they can. This documentary is the story of Momtaj Begum, who joins thousands of internal climate migrants streaming into the capital city of Dhaka. As the city buckles under the strain of the influx, the migrants are moved on by the government, which destroys their makeshift dwellings.

What's happening in Bangladesh is just the precursor for a scene that's likely to be replayed globally in the not-too-distant future.

https://www.youtube.com/watch?v=co5uywe-1Z8

Where Do We Go When the Seas Rise?

In this BBC World Service (2023) documentary, you'll be taken to a manmade island off the Caribbean coast to see how humanity might be able to adapt to rising seas. Another reporter heads into a high, cold desert to see if heading upward might not solve the problem. The documentary features conservation biographer Matt Fitzpatrick from the University of Maryland, who has drawn a map of how cities and towns are likely to feel 60 years from now and how to get a preview of your home's future climate. But we'll not be the only ones on the move.

https://www.youtube.com/watch?v=e0V1CtzFgrU

Arctic Sinkholes

Not far from the North Pole, the Arctic permafrost is melting, triggering massive methane eruptions that create giant sinkholes. The implications of global climate change are terrifying—but so are their effects on local villagers, who are losing their houses to the permafrost and must relocate. Where can they go? And what do the sinkholes mean for climate around the world? This intriguing documentary (Nova PBS Official, 2022) will provide insight into the world of tomorrow and the Arctic communities' response to this new threat to their lives and livelihoods.

https://www.youtube.com/watch?v=HvKpnaXYUPU

Key Chapter Takeaways

- Climate change is upsetting the lives of ordinary people. In many places in the US and Mexico, people are anxious and apprehensive. Where will the next wildfire erupt? Will they survive the next hurricane? Why did those wonderful city trees die? People are increasingly anxious, looking over their shoulders in case the next heatwave, wildfire, or hurricane should catch them off guard. Building resilient communities could help.

- Climate change is destroying cultures, communities, and social cohesion. People are moving, mostly within their own countries but occasionally across international borders, to escape the endless droughts, torrential floods, rising seas—and a tide of hopelessness. Yet the international community is dragging its feet, unable to legally define these new "climate migrants," let alone a way to protect those fleeing extreme weather. The world's richest individuals have emitted far more greenhouse gases than these people ever will, yet the wealthy are slow to develop real solutions.

- As temperatures climb, disease vectors like mosquitoes and ticks are colonizing new areas and bringing diseases, once confined to tropical zones, to Europe and the subarctic. Climate change is putting tremendous strain on healthcare facilities, as extreme weather increases doctor visits, emergency room visits, and even hospitalization. In the wake of climate-related disasters, healthcare workers and facilities are often stretched to their breaking point. The healthcare facilities of a world under climate change will need a new breed of healthcare workers, specifically trained to deal with natural disasters and difficult conditions.

- The economic fallout from climate change is affecting many sectors and industries, including tourism. Financial markets are fluctuating and changing, as investors move away from supporting fossil fuels; and businesses struggle to protect themselves from supply chain disruptions caused by floods and severe weather in other parts of the world, infrastructure and transport challenges, and rising costs. However, the changes the world is experiencing are also creating opportunities for green jobs in several sectors.

The first five chapters of this book have been crafted to enable you to understand climate change and appreciate the full implications of what we're up against in a climate change world. Now it's time to consider what can be done to reduce the pressure. There are solutions and technologies that can make the world a better, safer, more productive, and gentler place. We haven't reached the end game yet.

Mitigation Strategies–Big Picture Solutions

Picture a world where the rising tide of climate change is met with even stronger countermeasures. This is a world where nations come together and take meaningful action, ushering in an era of innovative solutions. In this chapter, we'll describe the pivotal strategies being employed on the world stage, showcasing humanity's commitment to healing our planet. Are we up to the challenge? Read on to find out.

The Paris Accord and Beyond

In December 2015, at COP 21, the historic Paris Agreement was signed. All signatories committed to reducing their nations' contributions to climate change by:

- significantly reducing greenhouse gas emissions to limit temperature increases to no more than 35 °F (2 °C), while working to reduce emissions even further so global temperatures would not increase by more than 34.7 °F (1.5 °C)

- reviewing their commitments every five years
- providing finance to assist developing countries with climate change mitigation and adaptation, while strengthening resilience (United Nations, n.d.)

The Paris Agreement is a legally binding international treaty that came into force in November 2016. Today, there are 194 signatories or parties (United Nations, n.d.)

Country-Specific Commitments and Progress

Nationally Determined Contributions (NDCs) refer to targets set by different countries to reduce greenhouse gas emissions and adapt to climate impacts. NDCs also indicate how a signatory plans to achieve these goals, finance them, and monitor progress to ensure their commitments are fulfilled. NDCs are updated every five years, becoming more stringent over time, so emissions are reduced faster, while adaptation measures increase. These plans enable countries to reduce emissions effectively in order to protect lives and livelihoods without delay.

Some countries have made ambitious pledges:

- Colombia aims to be carbon-neutral by 2050 and halfway to net zero by 2030. It is focusing on energy inputs in the agricultural, energy, and industrial sectors. Adaptation will be linked to national monitoring.
- Jamaica's tourist-based economy suffered tremendously due to the COVID-19 pandemic, yet it has increased its emissions cuts target to 60%. Plans include implementing efficient water use to combat waste and shortages.

- The Dominican Republic wants to transform its economy through climate action, and the private sector is providing significant funding for these initiatives. It is focusing on transport, one of its largest emitters, and plans to promote electric and hybrid vehicles.
- Morocco has intensified its actions and plans to further cut emissions by 46% by 2030. It has added nine new mitigation measures to the ones it already has. The country has 75% of global phosphate reserves and plans to mitigate manufacturing emissions. The country plans to switch to wind power to reduce emissions from desalination.
- Panama consulted with businesses across its major economic sectors and aims to restore 123,552 acres (50,000 hectares) of national forest. It has also increased its emissions targets to 11.5% by 2030.
- Rwanda was the first African country to revise its emissions targets and plans to cut emissions by 30% by 2030 across its entire economy. Rwanda has developed monitoring systems to track climate adaptation (United Nations, n.d.).

Financing Climate Action

Climate financing refers to finance used to mitigate climate change impacts and assist in developing and implementing adaptation measures. Global climate treaties, including the Kyoto Protocol and the Paris Agreement, require parties with more financial resources to assist those who have fewer resources at their disposal. Developed countries, which have historically produced higher emissions, therefore need to assist developing countries in implementing international climate change objectives. Developing countries need to ensure that the funding obtained is appropriate for their needs. Developed countries should also take the initiative in sourcing climate funding. Financial aid needs to be linked to emissions reduction and resilient development.

Climate funding mechanisms for developing countries include the Global Environmental Facility (GEF) and the Green Climate Fund (GCF). Two special funds have also been established—the Special Climate Change Fund (SCCF) and the Least Developed Countries Fund (LDCF). There is also the Adaptation Fund (AF), which was created in 2001 under the Kyoto Protocol.

In 2010, the COP established the Standing Committee on Finance (SCF), which assists it in enabling climate funding mechanisms.

Following the Cancun Agreement of 2010, developed country parties committed to work toward mobilizing $100 billion annually by 2020 to assist developing countries. The Paris Agreement confirmed this goal. A new goal will be set in 2025 (United Nations Climate Change, 2023).

Future Collaborative Agreements

Reducing emissions includes finding solutions and ensuring they are implemented. This requires the expertise of several sectors, including universities, industry, and government. Universities can inform, influence, and empower government, industry, and civil society to enable them to take action. By collaborating and designing solutions in association with government and industry, universities can help to ensure that cutting-edge solutions are implemented. Analysis is often required to create systems that will reduce emissions across several economic sectors, ultimately achieving net-zero emissions. Universities can assist with these aspects.

Because climate change is a global problem, international collaboration is essential. Across the world, universities are working through the Sustainable Development Solutions Network, focusing on net-zero research, education, policy implementation, and various forms of investment.

Transitioning to Renewables

Energy powers our industries, factories, and lifestyles, but we need to transition from fossil fuel sources. Fortunately, there are several renewable alternatives available, as well as new ways of providing and distributing energy.

Solar, Wind, and Hydro: Pros and Cons

These are today's premier renewable energy sources, which are powering homes, businesses, and communities.

Solar Energy

The amount of sunlight that reaches the Earth in one hour is enough to power the world for an entire year. Solar is an incredibly popular alternative energy source—in 2020, it accounted for 3.3% of total energy generated in the US. Solar panels last for around 25–30 years before needing replacing (Williams, 2023). As solar becomes more affordable, more people are switching to this form of renewable energy.

Benefits	Disadvantages
Lower emissions	Requires significant roof space in order to offset around 94% of your electricity bill
Affordable	Overcast weather can affect and reduce performance
Tax credits available	Might not be aesthetically pleasing for some homeowners
Sustainable	Rooftop solar is not portable, so it will need to remain in place if you move
Low maintenance	
Adds value to your home	

Hydropower

Hydropower is the oldest and most widely used form of renewable energy in the US, accounting for 37% of all energy used (Williams, 2023). It's created by using a water source such as a dam or even ocean waves to drive a turbine. This turns a generator that converts the motion into electricity. If a river or stream runs through your property, you can use this to generate your own power on site. You'll need a turbine, prop, and waterwheel or pump.

Benefits	Disadvantages
High efficiency—90%	Environmental and community impacts of dam building for large hydropower projects, including new infrastructure and displacement of people
Clean energy source	Flood risk
Having running water on your property will generate sufficient energy to avoid you having to rely on an outside source	Large reservoirs are costly and time-consuming to build—construction may take up to five years
At the community level, the reservoirs used for hydropower can also be used for recreation	Droughts may reduce the efficiency and effectiveness of hydropower
Dams used for hydropower also provide clean drinking and irrigation water, and control flooding	

Wind

Wind energy is best used in places that regularly experience high winds, including deserts, the American prairies, or off-shore sites. The wind turns turbines at the top of tall towers. These in turn spin a rotor that converts the motion into mechanical power. This feeds into a generator that makes it into electricity. Wind speed, air density, and cleared areas all determine the effectiveness of wind energy.

Benefits	Disadvantages
A clean energy source ideal for windy areas	Wind is unpredictable and this makes it difficult to determine how much electricity will be produced at any given time—low wind speed means no energy is produced
Low operating cost as wind is free	Turbines can be noisy
Wind turbines don't take up space and can be placed in an area that is not regularly used	Turbines can be eyesores due to their height above the ground
	Turbines can pose risks to birds and bats that might fly into them, especially at night or during migrations

Energy Storage Solutions

When installing renewable energy systems, it's often necessary to store the energy produced so it can be used when power is not actively being generated: for example, when the wind isn't blowing or the sun isn't shining. While household solutions are relatively simple, energy storage at scale is more challenging.

Energy can be stored by using batteries, flywheels, pumped hydro, compressed air, thermal storage, and hydrogen. Each solution has its benefits and disadvantages, with some technologies being more suitable than others for a particular application. Batteries are easy to use but degrade over time and have limited capacity.

Storing the energy generated by renewable systems has several advantages, including:

- balancing supply and demand
- reducing peak demand
- improving grid stability
- increasing resilience
- easing fluctuations in renewable sources such as solar and wind
- storing excess energy when available and releasing it when required
- maintaining the quality and reliability of the electricity supply
- providing backup during emergencies or other disruptions

(Ganyarpawar, 2023.)

Several factors continue to prevent the implementation of energy storage at scale. These include high costs, technical challenges, and a lack of incentive. Energy storage options may need to be integrated into an existing power grid and should be compatible with different platforms and protocols.

To create a sustainable energy future based on renewables, the development of large-scale energy storage systems will be crucial. Rising demand for renewables is expected to drive future development and implementation of these systems.

Decentralizing Energy Systems

Unlike the national power grid, decentralized energy systems focus on delivering power closer to where it is needed. This also increases the efficiency of renewable energy sources. Because they are so close to the user, these systems can generate both heat and energy. At scale, these systems could use combined heat and power plants. This further increases efficiency, as electricity generation often produces heat. Batteries, pumped hydro, and compressed air can all help to stabilize the grid by storing energy when supply exceeds demand, to release it to the grid during peak periods. Decentralized power systems can be connected to the grid (grid-tied) or operate independently.

Future Farms

Global food production will need to increase by around 75% in order to feed the entire human population by 2050. Making matters worse, around two-thirds of the world will live in water-stressed regions by 2025, so it will be more difficult to grow crops (Tiernan, 2019).

As mentioned previously, agriculture is particularly vulnerable to climate change and extreme weather events. However, there are ways in which farming can reduce climate change impacts. Organic farming focuses on improving the soil, which enables it to sequester more atmospheric carbon. Certain technological advances in recent years might improve farm productivity, while urban farms could supply fresh produce to city residents.

Organic Farming and Soil Health

Without healthy soil, humanity will not be able to meet its food, water, and natural resource needs. Healthy soils retain water, provide crops with nutrients, store carbon, and support beneficial microorganisms that promote nutrient cycling and decomposition. However, soils across the US are generally being degraded by unsustainable farming practices.

Soil stores around 80% of terrestrial carbon reserves, but poor agricultural practices such as tillage, plowing, and not maintaining soil coverage outside of the main growing season have led to vast amounts of carbon being released into the atmosphere. This has accelerated over the last 200 years, as more natural lands are converted to croplands or rangelands.

Research into organic farming has demonstrated that these farming practices can actually increase soil carbon levels, which contributes to climate change mitigation (Shade & Tully, n.d.). Farming practices such as growing cover crops, adding organic inputs like compost and manures, crop rotation (growing different crops on the same piece of ground at different times), and shallow forms of tillage can all help to improve soil health.

Precision Farming

In 2001, a tractor manufacturer installed GPS systems in all its tractors, which enabled farmers to precisely track their movements. As a result, farmers saw their tractor fuel bills decrease dramatically—by as much as 40% in some cases (Tiernan, 2019). This was one of the innovations that reflected the precision farming trend, where information technology, artificial intelligence, and agricultural engineering laid the foundation for precision farming. Using remote sensors, drones, and other technology, farmers gather vast amounts of data about their operations, from weather patterns to nutrient deficiencies in a single plant. This digital revolution is increasing yields while reducing inputs and pesticide use leading to a reduction in water and air pollution, and saving non-renewable resources like phosphorus.

Driverless tractors can be instructed to till only specific land areas, and quadcopters collect data on everything from soil chemistry to plant growth, enabling farmers to reduce irrigation requirements and increase yields. At present, many of these technologies require a certain level of digital literacy before they can be adopted. Some farmers who already have smartphones aren't always using available digital technologies to improve and streamline their operations.

Biotechnology

Biotechnology changes the genetic codes in crop plants to confer new traits such as improved growth, higher nutritional value, and greater pest and disease resistance. While not everyone is in favor of tampering with natural genetics, modern industrial agriculture is cultivating such crops on a small scale. However, biotechnology may be one of the tools that will improve food production in a climate-challenged world. With this in mind, scientists are focusing on developing crops that are able to tolerate drought, extreme heat, wildly fluctuating temperatures, and saline conditions.

The gene-editing bacteria CRISPR is being used to make photosynthesis more efficient and enable plants to draw nitrogen from the air, thereby reducing fertilizer use. This could also reduce groundwater pollution. One biotech company has created benign microbes that grow beside plants, encouraging them to use nutrients more efficiently, cope with environmental stressors, and increase yields. Natural predators and microorganisms are being harvested to help reduce pests and diseases. Farmers using tiny drone bees for pollination in areas where natural bees are scarce found that their yields increased by 20%–30% (Tiernan, 2019). Golden Rice, which has been genetically modified to contain vitamin A, is being rolled out in developing countries where children are deficient in this vitamin.

Urban Farms

As the amount of available agricultural land decreases, indoor urban farms are providing additional opportunities to grow food. Hydroponic and vertical farming systems are augmenting food supplies in urban areas, where they are supplying salad greens and other fresh produce. These systems don't use soil, are isolated from environmental stressors, pests, and diseases, and use drip irrigation. This means they rarely encounter the usual problems found on farmland.

Their main drawback is the amount of energy they consume to provide sufficient light for plant growth. Companies that manufacture light bulbs are already considering developing bulbs that transmit light at the right wavelength to enable photosynthesis, last longer and are more efficient. With the world rapidly urbanizing, such urban farms could become an important source of food for city dwellers in the future.

The Potential of Geoengineering

Technology might be able to reduce some of the worst effects of climate change in the future. There have already been discussions around carbon capture and storage, as well as reflecting sunlight back into space.

Carbon Capture and Storage (CCS)

While the effects of climate change are becoming increasingly apparent, world leaders are dragging their feet in implementing their NDCs, fossil fuel companies and other corporations are lobbying for business as usual, and the public has been slow to take up renewable energy, invest in electric vehicles, or even use public transport. This means that global temperatures could rise above the levels recommended by the IPCC. We need to find ways to reduce carbon dioxide in the atmosphere as quickly as possible.

Enter the world of CCS. The technology involves three steps:

1. Carbon dioxide is separated from other gases during industrial processes at coal-fired power stations, cement factories, or steel smelters.
2. This is compressed and transported via pipelines, trucks, or ships to another site for storage.
3. The carbon dioxide is then injected into underground rock formations and stored. Proponents of CCS technology envisage that the gas could also be stored in saline aquifers or oil and gas reservoirs, which are often a mile underground. In the UK, a massive off-shore aquifer dubbed Endurance, which exists a mile below the seabed, is being considered as a possible storage site (National Grid, 2023).

A related technology is carbon capture utilization and storage (CCUS). The idea is to reuse the harvested carbon in industrial processes to make plastics, concrete, or biofuels.

CCS is not new—it was first used in the US in 1972. Since then, natural gas plants in Texas have captured and stored over 200 million tons of carbon dioxide in underground reservoirs. By the end of 2022, there were 194 large CCS facilities worldwide, 30 of which were in use, while the others were still being developed. The industry says that CCS is a safe, proven technology that has been used for years (National Grid, 2023).

Reflecting the Sunlight

Solar radiation management (SRM) is another geoengineering technology that is being considered. The idea is to reflect incoming solar radiation back into space, either at the Earth's surface or in one of the upper layers of the atmosphere.

The SRM strategies that have attracted the most interest are:

- injecting sulfate aerosols into the atmosphere
- artificially brightening clouds over the sea (Scott, 2012)

Sulfate aerosols in the atmosphere would replicate conditions that prevail during volcanic eruptions, which would have a cooling effect on the Earth's surface. Cloud brightening would involve seeding clouds with tiny droplets of seawater to make them more reflective, mimicking the way sea ice reflects solar radiation back to space. Unlike CCS, which is a very costly exercise, SRM is touted as being quick and cheap. At this stage, geoengineering has yet to be embraced at government level.

Potential Impacts on Global Weather Patterns

Using technology to fix the climate problem is not necessarily the solution because the natural world consists of complex, interconnected systems. Using SRM could worsen ocean acidification because it might increase atmospheric levels of carbon dioxide. Spraying sulfur aerosols into the atmosphere could negatively affect the ozone layer. Geoengineering might not have the same effects everywhere. It could further alter precipitation patterns, for example, causing problems for farmers and the global water supply. Deploying geoengineering solutions would involve many actors, all with different interests. It is highly likely that there would be winners and losers, including the global climate.

Key Chapter Takeaways

- The Paris Agreement was a watershed in climate change discussions among world leaders and policymakers. Under the terms of this agreement, signatories were obliged to enter into NDCs—commitments to reduce their emissions. Climate finance vehicles were created so developed countries could fund developing countries' efforts to adopt mitigation and adaptation measures to cope with climate change. NDCs are reviewed every five years.
- Private individuals, communities, and utilities are investing in renewable energy solutions like solar, hydro, and wind power. All three have benefits and disadvantages, with solar power being the most popular with homeowners and hydropower being the most efficient. The difficulty with renewables is storing the power for use when it is not being generated. There are several options, with battery storage being more prevalent. For utilities, having decentralized energy distribution is beneficial as it helps to balance the grid, ensures that electricity is generated close to where it will be used and that the heat generated can also be utilized.
- The agricultural sector emits a significant amount of greenhouse gases. However, soil is capable of storing a considerable amount of carbon. By farming organically and improving the soil, farmers can increase the amount of carbon the soil holds. This feeds beneficial soil life,

improves the water-holding capacity of the soil, and makes it more fertile.

- Precision farming uses technology to gather vast amounts of information about farm inputs so that farming can become more efficient and increase yields. Biotechnology is creating plants that are disease and pest-resistant, grow better and are more nutritious. In urban areas, hydroponic and vertical farms are an option for providing fresh produce close to population centers.

- As the international community is dragging its feet when it comes to reducing emissions and mitigating climate change impacts, scientists are weighing in with technological solutions. These include CCS, which is already being done in many parts of the world. Carbon dioxide is captured from factories, smelters, and power stations and stored underground in rocks or reservoirs. SRM involves adding sulfur to the atmosphere to block the sun's rays and seeding clouds to reflect heat into space. Both these technologies could backfire, however, making climate impacts worse, and have not been endorsed by policymakers.

- While low-tech options can and are providing solutions to our climate dilemma, the rise of artificial intelligence (AI) is adding a new tool to our arsenal. AI represents a novel technological frontier and could be employed to provide new ways of reducing the impacts of climate change and its effects on the planet. There are indications that AI might be able to both predict and shape a sustainable future for humanity. If you find that a fascinating concept, then the next chapter will demonstrate how technology and ecology might converge.

Artificial Intelligence to the Rescue

In an age where climate change challenges often appear insurmountable, imagine a counterbalance—the vigor of human ingenuity—coming to the rescue.

Artificial intelligence (AI), once found only in the realms of science fiction, has emerged as an unexpected ally in the global battle against climate transformation. In this chapter, you'll find out how AI is being harnessed to reshape our approach to one of the planet's most pressing issues.

How Does AI Work?

Before explaining how AI is helping in the fight to slow climate change, it's necessary to grasp how this technology works. Machine learning is a function of AI—this enables computers to "learn" without being programmed. Instead, algorithms are used to match a mathematical model with a specific set of data through "training." An algorithm is a set of instructions for solving a problem or accomplishing a task. The more training data provided, the better the mathematical models become. The best-performing model is the one ultimately used.

The first step in machine learning is to collect as much data about a specific subject as possible. This may need to be over a very long period in the case of things like weather patterns or crop yields, and would necessarily include variables and human observations. Machine learning algorithms then "train" the model. Multiple models are often produced. Not all of the data is used. To see which model performs best, some of this unused data is employed to determine how well the learned model performs. After the best-performing model is selected, it can be deployed to provide information regularly.

Models can provide information either numerically (e.g. yield per acre) or categorically (e.g. low, average, or high). The results are only as good as the data that was used to generate them, so the better the data, the more accurate the results.

Predictive Abilities of AI

AI can help us prepare for extreme weather events by analyzing patterns and making predictions. Not only that, it can predict disasters and assist with disaster management, forecast crop yields, and more.

Enhanced Weather Forecasts

Have you ever listened to the weather forecast predicting a sunny day and then found it starts raining by mid-morning? Inaccurate forecasts are usually no more than an inconvenience. However, living in a climate change world means that an incorrect forecast could mean the difference between life and death, especially if severe weather is on the way.

Accurate forecasts are therefore becoming essential and this is where AI could assist. Two AI systems were tested in July 2023. Pangu-Weather reliably predicted global weather up to a week in advance, while NowcastNet correctly forecasts rainfall up to six hours ahead of time (Solis-Moreira, 2023).

AI's potential lies in its ability to forecast not only daily weather but also extreme weather events. Weather forecasts in most countries are based on mathematical equations that create computer simulations of the oceans and atmosphere. The accuracy of AI forecasts is enabled by more advanced and powerful computers, improvements to computer weather models, and more data. To train the Pangu-Weather model, researchers loaded the system with 39 years of global weather information. Pangu-Weather produced accurate forecasts 10,000 times faster than conventional weather models but was unable to predict precipitation levels (Solis-Moreira, 2023).

For that, they needed NowcastNet. This AI system delivers detailed precipitation forecasts focusing on heavy rainfall events in local regions. Radar information from the US and China, together with deep learning, enabled the system to predict precipitation over 1.6 million square miles of the US up to three hours in advance. Meteorologists who tested the system found it to be accurate 71% of the time (Solis-Moreira, 2023).

Although these systems are promising, they won't replace human meteorologists and current weather prediction models for around 5–10 years because they can't generate physics-based mathematical equations (Solis-Moreira, 2023). To be truly accurate, these AI forecasters would need to perceive the way in which moisture, air, and heat move through the atmosphere. In addition, if the AI is trained on low-quality or inaccurate data, then the forecast could be less accurate in some regions.

Early Warning Systems for Natural Disasters

Extreme weather often results in disastrous outcomes for those in its path. Especially for those shoring up their businesses or navigating clogged highways when under evacuation orders, it might feel that warnings could have been given earlier. In March 2022, the World Meteorological Organization (WMO) undertook to explore the potential of AI to improve global disaster mitigation.

Natural hazard models have improved tremendously over the last few decades and their warnings have saved numerous lives. Nevertheless, there is considerable room for improvement, as one in three people are still not adequately protected by early warning systems (ITU News, 2022).

Data combined with machine learning can analyze past events and establish patterns to improve forecasts. AI algorithms can identify climate patterns, at-risk areas, and populations, and even send early warnings. They can foretell the outcome of a natural disaster and predict the financial cost for specific countries and populations. Even agencies with limited resources could develop disaster management systems based on accurate information.

Satellites and other meteorological equipment would still be used but AI could augment this by creating models that indicate the impacts of extreme weather on populations and ecosystems.

AI could also be used in the following ways to facilitate better disaster management:

- monitoring and early detection of potential risks
- predicting and projecting likely weather events
- communicating what has occurred after the event (ITU News, 2022)

Cyprian researchers are finding ways to use remote sensors, sometimes attached to drones, and machine learning to enable objects to be tracked from the air. Other scientists at Lancaster University in the UK have developed a disaster-mapping and damage-detection system that allows teams to prioritize certain areas when providing relief.

Using international geological information, AI could improve earthquake and tsunami warnings. Wildfires could be studied and data analyzed to enable early detection so communities can be moved out of harm's way. New malaria software predicts the appearance of new water bodies where malaria-causing mosquitoes may breed (ITU News, 2022).

AI for Conservation

The role of technology has changed the conservation landscape forever. With its ability to sift through masses of data and "learn" from them, AI offers endless opportunities to further the aims of conservation in several areas. It is already enabling conservationists to monitor rare species and habitat change, predict species movements and novel threats, and find out more about rare plants and animals.

Habitat Monitoring and Restoration

Climate change is altering habitats so that wildlife, birds, and other creatures are colonizing new areas. This can both reduce and exacerbate threats, especially for species under pressure. This is adding to the complexities of conserving unique and important landscapes, together with planetary biodiversity. AI makes it easier for conservationists to establish how habitats are changing by combining satellite tracking with machine learning. This makes it easy to observe land-use changes that could affect local and regional wildlife.

AI enables the creation of restoration projects that can withstand changing weather patterns and climate shifts. Conservationists need to continually evaluate and monitor restored habitats and adjust their management strategies accordingly. To do this efficiently, scientists need to constantly collect and analyze data, which is labor-intensive and time-consuming. AI can do much of this work quickly and effectively, freeing up man hours and ensuring that any management interventions are effective.

Species Population Tracking

Technology is enabling conservationists to track wildlife efficiently and safely. Previously, wildlife monitoring involved tagging animals and tracking them on the ground. Today, camera traps and AI-powered drones are enabling researchers to monitor animals from a distance without disturbing them. Using this technology, it is possible to identify and monitor individual animals, while also observing behavioral changes or health problems. Because of its ability to process vast amounts of data, AI could find patterns that human observers don't see when they are in the field.

Satellite imagery, devices attached to animals, and camera traps all provide copious amounts of data that can be analyzed by AI. This enables conservationists and researchers to monitor things like changes in population sizes, shifts in habitats, or animal and bird migrations. AI can identify changes in bird migration, which can signal altering weather patterns, for example. When population sizes change, this can indicate alterations in their habitat due to climate change effects.

In Africa, AI was effectively used to track elephant populations. It analyzed satellite images to count and map the animals, providing valuable insights into elephant numbers and distribution across the continent.

Making Predictions

One of the particular benefits of AI for wildlife conservation is that, once it has been trained, it can sift through copious data sets and make predictions. AI quickly and effectively predicted bird migratory patterns across an entire continent. The job would have taken around ten years of computing time to process if AI had not been used (IUCN, 2023).

AI is helping to make predictions about habitats and species about which little is known. Kew Gardens in London is using AI to establish which of the world's plant species are most at risk of extinction. Software powered by AI took just one day to establish the main threats faced by over 47,600 plant species (IUCN, 2023).

Other AI projects are using images uploaded by the public to identify plants and animals. AI then uses this data to establish global biodiversity and monitor how climate change is impacting different species. AI is also analyzing thousands of satellite images to create detailed forest maps that show the species present and their carbon storage capacities while determining the overall health of the forest. This is important for forest management in an age of climate change.

Anti-Poaching

AI is also being used in global anti-poaching efforts. The Protection Assistant for Wildlife Security (PAWS) uses data about a particular protected area, as well as previous anti-poaching patrols and their effectiveness to identify the best patrolling routes for rangers.

The University of Southern California is using AI to forecast poaching hotspots by analyzing poaching data to establish patterns and trends. This helps conservationists and authorities to establish which areas are likely to be targeted by poachers so that proactive steps can be taken.

Limitations of AI in Wildlife Conservation

While AI can be a powerful tool, it is essential to ensure that the models are trained on the correct data. This was clearly demonstrated when a wildlife monitoring tool found giraffes in the frigid Canadian city of Edmonton!

To use AI technologies effectively and ethically, the following guidelines should be followed:

- Data quality should be relevant to the subject at hand and deliver the desired outcomes.
- The potential for harm should be considered and mitigated.
- Issues around privacy and surveillance should be addressed.
- The process should be explainable, transparent, and fair.
- The use of AI should balance the rights of wildlife, individuals, and communities.
- Users should know when it would be inappropriate to use AI and avoid doing so (IUCN, 2023).

Efficient Energy Management with AI

AI can enhance energy efficiency in both conventional power systems and renewable energy installations. Many buildings have lights that switch on when someone enters an office and switch off within a few minutes of them leaving. This can save a considerable amount of energy. These systems are powered by AI.

Smart Grids and AI

US power plants were not originally built to interface with renewable energy systems. When electricity demand surges and there is insufficient supply, utilities switch on backup power plants powered by fossil fuels, which is expensive and wasteful. Enter smart grid technology. Smart grids are fully automated electricity delivery networks that monitor and control the flow of electricity to every consumer.

For smart grids to work effectively, electricity needs to be produced in many places (distributed generation), rather than being produced centrally and distributed over long distances. If energy is produced in several places, power can be diverted automatically from another generation facility should there be a supply interruption. Consumers are not even aware that this is happening, as their power supply remains constant.

The shift to a decarbonized economy is likely to disrupt conventional electricity supply processes. Voluminous amounts of data are required to operate a smart grid system and this is where AI comes to the fore. Properly trained AI would be able to analyze data such as demand, location, weather, and available generation equipment. Thereafter, it will be able to decide exactly where to source the electricity for each home connected to the smart grid and determine the energy costs thereof. AI can also be proactive.

If the system knows it is going to rain for three weeks, for example, it can scale up production from other sources to allow for losses in solar generation at that time.

Consumers could also be connected directly to the grid, so that you could automatically do your laundry during off-peak hours, when power costs less, for instance. Smart metering monitors every appliance and outlet so homeowners receive detailed breakdowns of how much energy their home is using and what items are raising their consumption. They are therefore able to make informed decisions about their energy usage and see where they can save.

Optimizing Renewable Energy Sources

More people and businesses are installing renewable energy and the trend is set to continue as the realities of climate change set in. These power sources need to be managed efficiently and optimally—and this is where AI can assist.

AI algorithms are able to accurately predict the electricity output from wind farms and solar arrays. This enables utilities and other energy companies to optimize their production schedules, so there is less waste and better efficiency. Sensors on solar panels, wind turbines, and other renewable energy infrastructure deliver data that AI systems use to automatically adjust solar panels and wind turbines so that they deliver optimal amounts of energy. This improves energy generation. While reducing maintenance requirements and costs, ensuring that the system is more durable and reliable.

Reducing Energy Consumption

AI is a useful tool for reducing energy consumption. Consuming less energy reduces both fossil fuel usage and greenhouse gas emissions. Globally, buildings use about 40% of all electricity produced (Webb, 2023). Much of this is lost due to inefficient systems and human behavior. Sensors and other hardware in buildings can deliver information about energy use to AI algorithms, which identify usage patterns. Building owners and managers can then make informed decisions about optimizing and reducing energy use. AI automatically adjusts heating, ventilation, and air conditioning systems throughout the building to use energy more efficiently.

Autonomous vehicles may be able to reduce transport emissions and fuel usage by choosing more efficient routes, relieving congestion, and facilitating improved traffic flows.

AI can be used on production lines to optimize production, enabling manufacturers to reduce energy use and costs, while also lowering emissions.

AI's Energy Footprint

One of the great ironies is that, for all the energy-saving potential of AI applications, the technology itself gobbles up vast amounts of energy. Supercomputers that run AI programs are powered by the regular electricity grid and sometimes need backup diesel generators to supplement the supply. AI technology actually emits as much carbon as the global aviation industry.

There's no easy answer to making AI more energy efficient. One option is to use rechargeable lithium batteries. However, the extraction of this rare earth mineral requires 500,000 gallons of water per ton of lithium extracted (Jones & Easterday, 2022). In Chile, which holds significant lithium resources, mining companies and communities are clashing over land and water rights. The US government is working on developing solid-state lithium batteries, which last longer and have better performance. New kinds of batteries are also being considered, with the electric vehicle industry investing significant R&D into developing new battery technologies.

Some data farms have opted to use other forms of renewable energy like geothermal or hydroelectric sources. Iceland has an abundance of both and has become a magnet for new data centers. Another benefit is that no cooling is required, as Iceland has a naturally cold climate.

Ethical Implications of AI

Writers, artists, and songwriters are expressing concerns about AI, as consumers gasp in amazement at paintings created by algorithms and lawyers scramble for ways to protect their clients' copyrights. Yet it is not only the creative arts that are being disrupted by AI. AI has the potential to invade your privacy, use your data in unexpected ways, and even skew a particular data set to your detriment.

Data Privacy Concerns

The hijacking of information on social media to fuel a US presidential campaign made headlines in 2016. Around the world, people began questioning whether tech companies had a right to use the personal information we post on social media in this way.

Privacy is a basic human right. This means that you have the right to keep your personal information confidential and prevent others from obtaining it without your consent. Your personal information is yours to control and distribute at your discretion. Third parties may not interfere with your personal or business relationships or spy on you.

The trouble is that AI algorithms are becoming more efficient and might be able to unintentionally violate an individual's privacy by finding a subtle pattern in a data set of which no one is aware. AI also acquires extensive personal data that can potentially be accessed by criminals, cyberbullies, and other questionable individuals.

AI Bias and Environmental Decisions

It is important to remember that AI is not an autonomous entity but a system created by humans. Although relatively powerful, these systems are still subject to the flaws inherent in human nature, including bias.

If an AI system is trained on biased data, the resulting models will reflect this bias and may make or inform discriminatory decisions. AI systems obtain personal data from online activity, social media, and public records, all of which can indirectly provide a significant amount of information about you. This can be used to reinforce bias and discrimination. An example is an AI system biased toward men that is used to screen applicants for a job interview. If a woman applies, their application will automatically be omitted, effectively discriminating against them.

Because AI models use copious amounts of data, they rapidly acquire all the facts about a particular situation. This is where transparency and recognition of bias come in. Creators of AI need to ensure that the decisions their systems make are ethical. Among other things, this means anticipating potentially adverse effects and rooting them out. Any unintended bias in the data should then become obvious.

Training data for AI will affect the solutions it devises, in much the same way as human actions are predicated on our world view. The AI system will need to be able to evaluate which decisions are moral and ethical and be able to weigh up the consequences of its decisions. When supplied with ethical guidelines, an algorithm may be designed so the AI can develop actions associated with ethical values like human rights, justice, and the common good. Well-trained AI models are reliable and unemotional. These systems should be programmed to be impartial and consider all the ethical ramifications of a decision before selecting a particular course of action. While AI systems are powerful tools, it will be necessary to evaluate a model's suggestions before determining its trustworthiness.

AI and Environmental Decision-Making

AI could either help to protect or further devastate the natural environment depending on whether the models it uses are geared toward protecting or abusing resources. While AI can be helpful, it has also been shown to have drawbacks, especially with regard to environmental justice—ensuring that environmental laws and policies are enacted and implemented fairly to the benefit of all people everywhere.

Below are a few difficulties that have been experienced when using AI to develop systems to address environmental issues:

- If the training data is incorrect, this could create an unintentional bias. For example, if the model is trained on data for a certain number of years but tested on data for different years, then it might not properly account for climate change.

- Another difficulty arises when data is based on information derived from sensors. Many sensors require sunlight to operate properly and nocturnal observations are therefore omitted. Satellites using passive sensors also need clear skies to deliver accurate information.
- Different human populations and regions should be included in the data used for the training, validation, and testing of AI models. National radar networks don't cover certain areas completely, for example, so some populations might be under-represented in a particular data set, resulting in poor decision-making.
- Data from sensors could be incomplete, due to either sensor errors or missing sensors. Sensor placement might be limited in inaccessible places such as mountaintops. Additional sensors might then need to be deployed. Alternatively, AI could estimate sensor values in places where they are missing or compromised. Economic considerations might also affect sensor placement, so there could be more sensors in affluent areas.
- Environmental models usually consider many processes that happen in various places at different times, so it is difficult to find AI models that represent all potential processes. This means that some aspects might be neglected.
- Data for environmental predictions is naturally biased in favor of geographical areas with higher human populations, as there are usually more records and data for these areas. This could lead to AI over-predicting extreme weather events in areas with higher populations while under-predicting them for areas with lower populations and less data.

- AI models can be hacked and false data inserted. This can perpetuate fraudulent activities, such as bogus insurance claims.
- The weather may cause poor AI outcomes when sensors are unable to operate due to power outages caused by storms. Sensors might be damaged or destroyed during severe weather.
- Data used to train AI models affect every aspect of the model. There are numerous choices scientists make when selecting appropriate data, which can significantly influence results. There are also different AI models and model choices can affect outcomes. The choices a scientist makes could lead to unexpected and inaccurate outcomes with potentially severe consequences.
- AI models generally find patterns in the data they are given but there may be unusual correlations that are not immediately obvious. AI might then produce solutions that are not feasible in the real world. One of the ways of determining whether the AI has developed a faulty strategy is to ensure that the models can be understood by human experts (interpretable or explainable AI). This means that potential problems can be resolved prior to implementation.
- Some advanced types of AI can generate images that look authentic but lack accuracy. In one study, it was found that experienced forecasters relied on their own interpretation of the facts rather than simply accepting the compromised images supplied by AI. These methods are sometimes designed to maliciously trick people into believing misinformation and fake news.

(McGovern et al., 2022.)

AI Regulation and Oversight

ChatGPT exploded into our lives in late 2022, bringing the potential of AI into public consciousness. "It took one day to write code that would have taken me weeks," a friend of mine enthused. As we have seen in this chapter, AI isn't always going to be a boon to humanity and will very likely be subject to the same flaws as human beings because we're designing the models and training them on the data sets we select.

Some sources suggest that AI could become one of the world's most highly regulated industries because of the dangers lurking beneath its innocent I of high-speed efficiency. China has already enacted laws for managing AI, reining in some of its more dubious uses.

The EU is currently negotiating the AI Act, a process that began in 2021, well before anyone knew about ChatGPT (Hays, 2023). The legislation aims to categorize AI according to its risks, outlawing high-risk, unacceptable uses. High-risk AI includes that used for biometric identification, worker management, education, the legal system, and law enforcement. Some AI tools would need to be assessed and approved before release.

When it comes to generative AI—the type used for applications like ChatGPT and Open AI—AI-generated content would need to be disclosed, together with details of the data used to train the models. In anticipation of these laws, technology companies have stopped revealing their training data sources. Companies would need to demonstrate how they had mitigated legal risks before releasing any AI tools or models. All foundational models would need to be registered in a central EU database.

Brazil appears to be following the EU approach, focusing on human rights. Legislation may hold the creators of AI models liable for any high-risk AI system that causes harm.

The US is still lagging, with the government having released certain executive orders and obtained advice from tech industry experts and company CEOs. However, progress has been slow, although American lawmakers are unanimous in their opinions that AI needs to be regulated. Following the release of regenerative AI, the US copyright office has indicated that it is considering implementing federal rules due to the implications of this type of AI on creative industries. Generative AI is widely recognized for its ability to create new content that is similar to, but not identical to, the data they were trained on whereas regenerative AI is not a standard term and could theoretically refer to AI systems with self-improving or self-restoration capabilities.

The UK has been vocal about becoming an AI superpower and has no intention of regulating it at this stage, having decided to assess the situation as AI develops. The idea is to avoid constraining businesses that wish to use the technology.

China is leading the world in AI regulation and has already enacted legislation concerning recommendation algorithms. These will soon be extended to include deep synthesis or deep fake technologies. Regenerative AI is next. AI developers will need to ensure that their training data and models are accurate.

AI in Action: Real-World Case Studies

In this section, you'll find case studies to demonstrate how AI is being used to reduce climate change causes and impacts.

Modeling Urban Microclimates

Context

Urban areas are particularly sensitive to heat waves and rising temperatures, as concrete, glass, and other solid surfaces tend to absorb heat, making cities hotter and drier than their rural counterparts. Urban planning needs to mitigate these heating effects on the local microclimate. Planners must visualize how the wind circulates in cities, blowing down streets and moving around structures. They use computer simulations to provide this data.

The Issue

These models are costly and time-consuming to run, which makes it difficult for planners to compare and evaluate different scenarios.

AI Solutions

An Austrian technology company has created an AI system that uses deep learning to rapidly predict the results of these often complex simulations. This enables urban planners, architects, developers, policymakers, and municipalities around the world to easily access information on how new or existing urban plans might affect the local climate (AI and the F6S community, 2024 Feb 8).

Outcomes

The prediction models can be aligned to municipal or governmental planning platforms, so the potential outcomes of different designs can quickly be assessed and the best options chosen.

Improving Steel Recycling

Context

Steel manufacturing releases around 20% of greenhouse gas emissions, which are generated during mining and transportation (Global Partnership on AI, 2021). Recycling reuses steel from several sources, reducing the industry's reliance on raw materials.

The Issue

The difficulty is that each batch of scrap steel has different chemical compositions, which affects the strength and reliability of the new steel produced. Manufacturers add freshly mined materials (alloys) to each batch to ensure that the steel is of high quality. To avoid negative outcomes, many manufacturers tend to add more alloys than necessary, resulting in higher emissions.

AI Solutions

A US company has partnered with a major recycled steel manufacturer to reduce the amount of alloys added during production. The AI software learns from historical data to determine the amount of new material that needs to be added to a batch of molten steel. This frequently reduces the amount of new material needed. As the AI system does this in real-time, this also reduces the length of time that the steel needs to remain molten, cutting energy usage.

Outcomes

Over the past three years, AI software consistently reduced alloy use by up to 34% at five steel plants. This has prevented some 450,000 tons of carbon dioxide emissions a year. If this approach is adopted across the industry, the technology could prevent around 11.9 million tons of carbon dioxide emissions annually (Global Partnership on AI, 2021).

Climate Risk Finance

Context

As mentioned previously, climate change is likely to have a tremendous detrimental impact on global stock markets.

The Issue

Without adequate data, It is difficult to predict what these will look like and to inform investors. New legislation in North America and Europe has created a significant data set about how climate risks are likely to affect corporations and businesses that are publicly traded.

AI Solutions

AI algorithms process large amounts of text, identifying key passages in copious reports by using Natural Language Processing (NLP). Sustainability experts, who currently do these analyses manually, supply relevant data for training models. NLP models can be used either independently or together with human analyses. This generates information quickly.

Outcomes

Users of NLP technology in Canada and Europe isolated the climate risks mentioned in the annual reports of numerous companies. Besides informing potential investors, this has revealed that some industries, such as energy provision, are more open about disclosing their potential exposure to climate risks than others. (Task Force on Climate-Related Financial Disclosures)

Key Chapter Takeaways

- Artificial intelligence (AI) uses machine learning to train mathematical models to analyze copious amounts of data, forecast trends, and make predictions. Recent innovations have shown that AI models can use historical trends, satellite imagery, and other data to provide more accurate weather forecasts, as well as predict heavy rain days in advance. AI can be deployed to deliver early warnings of extreme weather events, giving disaster management ample time to make preparations, evacuate people, and protect infrastructure.

- AI can be used in nature conservation, providing accurate data on species movements and behaviors, particularly with regard to climate change. AI has enabled researchers to accurately forecast bird migrations in a climate change world, for example, and show how altered weather patterns are affecting the survival of thousands of plant species. AI has been used to inform on habitat restoration and evaluate management decisions.

- Electricity management can become more effective using AI. This technology automates and improves the functioning of smart grids that deliver electricity directly to end users, which makes generation more efficient. Households and businesses are able to evaluate their energy consumption when AI provides them with accurate information about their usage patterns. This reduces emissions.

- Nevertheless, AI is still in its infancy and there are certain drawbacks. There have long been privacy concerns around AI and these are likely to continue. It is important to ensure that AI models are trained on the correct data set to avoid unexpected outcomes, irrelevant data, or unintentional bias. Using sensors to obtain data for AI models can be problematic, as they may be absent due to the terrain, be insufficient in number, or deliver poor data after dark. This can affect the quality of data produced and the decisions made by AI systems.

- Ethical, responsible AI use is essential to avoid negative outcomes, unintentional bias, and data being used to perpetrate fraud and criminal activities. China has already implemented laws concerning the use of regenerative AI and further legislation is in the pipeline. The EU and Brazil are designing legislation based on the risk potential of AI, with Brazil also including human rights protections. The US has issued executive orders pertaining to AI development and is working in association with tech companies, although no legislation has been enacted as yet. The UK has chosen not to regulate AI at this stage to enable the technology to develop further and allow businesses to use it as they deem fit.

- The climate crisis is becoming so concerning that it is not enough to sit back and allow technological fixes such as CCS, sulfur aerosols, cloud seeding, and AI algorithms to reduce climate change impacts and protect us from disasters. You need only analyze your personal carbon footprint and that of your household with an online calculator to see that we ordinary individuals are contributing to emissions and climate change through our lifestyle choices.

- The situation outlined in this book may seem depressing. The good news is that you can personally make better decisions that will contribute to global change. There are many things you can do to reduce your emissions and mitigate climate change impacts. In the next chapter, you will discover actionable, everyday solutions and learn how every individual on the planet can fight climate change.

Empowering the Ordinary— Everyday Actions with Global Impact

Picture a world where several small actions ripple outward, creating waves of change far beyond their origin. While grand gestures and initiatives grab the headlines, it's the choices of ordinary people that hold the keys to transformation. In this chapter, you'll discover the surprising power of everyday decisions to shape the future of the planet.

Choices for Sustainable Living

Many of the swaps needed to live in a more eco-friendly manner don't take long, while others may require some forethought, expertise, and financial outlay. Some are easy, like changing your lightbulbs. Others might be more difficult but you'll know that you're doing your part to save the planet.

Creating an Eco-Friendly Home

There are simple hacks you can implement to live lighter on the planet and reduce your carbon footprint. Many of the suggestions below relate to energy usage, a prime driver of emissions, especially if you are using electricity derived from coal-fired power stations. As you continue on your sustainability journey, you will find countless other ways to reduce your footprint but here are a few ideas to get you started:

- Switching to LED light bulbs is a simple swap that makes a big difference. They last longer, provide more light, and are more energy efficient than other bulbs.
- Traditional water heaters consist of a tank full of water that is continuously stored and heated 24/7. This uses a significant amount of energy. Installing a tankless water heater means that the hot water you need is heated on demand as it moves through your plumbing system. Not only will this reduce your utility bills, it will ensure that you always have plenty of hot water.
- A surprising amount of household heat literally goes out the window—around 25% to be exact. Upgrade your windows to dual panels that are properly sealed to maintain your home's temperature at the lowest cost. You may be able to obtain tax credits or rebates to help with the cost of installation.
- Another benefit of technology is the development of more energy-efficient appliances. When buying new appliances, look for ones that are Energy Star certified. You can also optimize energy use by:

  o   running dishwashers and washing machines only when you have a full load
  o   using smaller appliances rather than larger ones, like toaster ovens
  o   covering pans while cooking

- o   switching off appliances and devices rather than
      leaving them on in standby mode as this still
      consumes electricity
- o   unplug so-called vampire electronics like TVs, game
      consoles, computers, microwaves, toasters, alarm
      clocks, cell phone/laptop chargers, curling irons,
      and cable boxes that all consume energy even when
      turned off.

- Install a ceiling fan rather than using air conditioning to
  reduce your energy use and lower your utility bills.
  Energy Star-rated ceiling fans are 60% more efficient
  than conventional ceiling fans and you'll reduce your
  emissions too.
- Consider switching to solar if you live in a sunny
  location. Switching to solar isn't cheap but the energy
  savings over ten years are well worth the outlay. You'll
  reduce your emissions as well.
- If your home's plumbing dates back to pre-1994, you
  might be using old technology that requires significantly
  more water than modern fittings. Low-flow showerheads
  and water-sensing toilets will reduce both water and
  utility bills (Glover, 2022.).

Eating to Save the Planet

The way we eat has a tremendous impact on natural resources. When you purchase food, go for organic, locally produced options that are in season. Buying organic supports organic farmers, who use natural techniques and protect the soil so it sequesters more carbon. By buying locally produced food in season, you reduce the emissions from transport and refrigeration.

Another way of saving the planet through diet is to reduce your meat consumption. Raising animals requires more resources, including land and crops used for animal feed, as well as the energy used during harvesting and processing. Start with Meat-Free Mondays and go from there. Opting for pasture-raised meat that has been farmed locally will support local farmers and reduce emissions.

Food waste is a considerable problem in the US, with large quantities of food being wasted at every stage from the producers to the home. To avoid throwing it away, never buy more food than you can eat. This waste food ends up in landfill sites, which release methane, a potent greenhouse gas, into the air when the waste decomposes. Alternatively, store food properly by freezing or preserving it.

By composting your organic waste, you reduce the waste your household generates and you can use this valuable soil amendment in your garden. Vegetable and fruit peelings, rotten fruit, used coffee grounds, tea bags, and eggshells can all be composted together with garden clippings. Avoid composting cooked food and bones, and never put garbage or cat litter into your compost heap.

When eating, opt for reusable dishware, flatware, cups, bottles, and containers. Disposable, single-use items contribute to your household's carbon footprint. Many items made from plastic are derived from fossil fuels and take years to decompose. The energy used and emissions created by manufacturing and transport also add to the footprint of these products.

Become a Conscious Consumer

Did you know that your buying habits can fuel sustainability? Being a conscious consumer means becoming aware of your purchasing decisions. What you buy determines your environmental impact. Buying fewer items reduces emissions and facilitates ethical purchasing decisions.

Start off by choosing to buy less. Only purchase essential items and avoid making frivolous, impulse purchases. Take care of your possessions so they last longer, and repair usable items rather than throwing them away. Aim to live simply and take things slowly.

Purchase quality products that can be reused. In this way, you can contribute to the circular economy by fostering recycling and reducing natural resource use. Buy local and support businesses in your area, especially if they have a low environmental impact.

Become more community minded. Buy secondhand, reuse, and swap items with friends and family members.

Community-Led Initiatives

Living more sustainably means becoming more community oriented and working with others to save the planet. There are several ways in which you can get involved in community work.

Local Conservation Projects

Community conservation refers to conservation efforts that involve local communities. Scientists work with residents to save local species or conserve important natural resources and landscapes. These interactions often start with honest conversations, where conservationists discover the community's needs and aspirations. In turn, the community learns about the benefits they may derive from conservation, such as better living conditions or tourism opportunities.

People need a livable planet that provides them with clean air and water, and healthy, resilient economies. Aiming to conserve 30% of America's wild places over the next ten years, President Biden launched the America the Beautiful initiative in 2021. This aims to redress climate change, biodiversity and natural land loss, and unequal access to wild places.

Communities across the US are working to conserve special wild places closest to them. There are numerous local conservation and restoration efforts taking place across the nation that are helping to fight climate change and environmental justice, strengthen our economy, conserve wildlife, fish, and birds, and improve outcomes for local people.

America the Beautiful aims to:

- create more parks and safe outdoor spaces for people in communities far from natural spaces
- create jobs through projects aimed at promoting resilience and restoration
- support tribal conservation and restoration initiatives
- incentivize and reward voluntary conservation efforts being made by farmers, ranchers, fishermen, and forest owners
- expand the collaborative conservation of wildlife habitats and corridors

For decades, a broad range of people have been involved in stewarding the land and water close to their communities and neighborhoods. It has become clear that the most effective conservation strategies are those that tie into the needs and priorities of local families and communities.

Community Gardens and Urban Greening

There are two types of gardens in urban areas. Conventional urban gardens are simply those that are planted in cities or suburbia, while community gardens are laid out in public places, often on community land. The latter are normally shared by everyone in the community. They may consist of individual plots where every person has their own space and maintains it, or they can be collaboratively managed by the community, with everyone contributing to planning, plant care, and harvesting. Some community gardens are a combination of the two. These gardens arise out of the need to green the neighborhood, restore degraded landscapes, or revitalize vacant lots.

Community gardens have many advantages, including:

- increasing property values
- increasing the productivity, aesthetics, and safety of vacant land
- producing food for the neighborhood, lowering grocery bills for residents
- increasing biodiversity and supporting pollinators
- introducing a variety of plants to previously denuded places
- absorbing stormwater, sequestering carbon dioxide, producing oxygen, and providing shade
- generating social opportunities, including cross-cultural interactions

- providing recreational and educational neighborhood opportunities
- bringing nature into urban areas
- enabling locals to connect to nature and their environment, creating positive mental health
- avoiding dumping, safety hazards, and criminal activity on vacant lots
- helping residents feel safer (Land Stewardship Center, n.d.)

Grassroots Movements for Policy Change

A grassroots movement encourages individuals to take action to influence social and political issues. They are considered to be bottom-up approaches that can have considerable influence, as they move from being small discussion groups to global networks in some cases. Numerous people are needed for grassroots movements to flourish and become effective in achieving shared objectives.

These campaigns raise money for their causes, increase public awareness of certain issues, and aim to further political participation. Social media networks have become excellent tools for grassroots activists to promote their causes to a wide audience.

The failure of governments and world leaders to make the commitments that would help to mitigate climate change by reducing emissions has frustrated the public—but it has also mobilized grassroots campaigns.

The Keystone XL pipeline was intended to bring oil from the Alberta tar sands in Canada to US refineries for processing, carrying some 830,000 barrels of crude oil to Nebraska a day. Construction was due to start in 2008 at an estimated cost of $9 billion. However, environmental groups, landowners, North American tribes, and climate change activists began a vigorous opposition campaign. The fight continued for 12 years, delaying construction. A government permit was approved in 2017 but the pipeline continued to face legal challenges and delays. Finally, in mid-2021, President Biden revoked a key permit that would enable the pipeline to cross state lines. TC Energy, formerly TransCanada, the pipeline's owner, incurred losses of $1.8 billion on account of its suspension (Reuters, 2021).

Extinction Rebellion (XR) was established in the UK in 2018 with the aim of using nonviolent activism to compel government action to prevent climate tipping points, together with associated biodiversity loss and societal collapse. They started by occupying Greenpeace's London offices, followed by a public launch outside the UK Parliament. Several XR activists are willing to be arrested and imprisoned for their actions, with mass arrests being one of the group's tactics to impact police time and create disruption. They are inspired by the Occupy movement, Indian salt marches, suffragettes, the civil rights movement, and Eastern European democracy movements. Several arrests were made in the UK during the April 2019 protests, including prominent politicians from the Green Party and some opposition party leaders (Wikipedia Contributors, 2023d).

Education and Advocacy

There are several ways in which we can educate school students about climate change, as well as actions they can take to change the status quo. Several advocacy groups are involved in encouraging climate change-related discussions, raising awareness, and aiming to change behaviors.

Engaging in Community Dialogues

It is difficult to talk about climate change because the subject is so complex, with far-reaching implications. Misinformation and denialism often obscure the facts, making it even more challenging to engage with others on the topic. Many people also prefer not to make causal links between their lifestyles and climate change and tend to avoid having open, honest discussions because of its implications. Lack of understanding and denial can effectively prevent climate action from taking place.

All these factors make it imperative to provide the public with better information using both traditional media and online options like websites, blogs, and documentary videos. However, there is no evidence that providing more information leads to a deeper understanding of the issues, let alone inspiring action or causing changes in behavior.

Climate conversations work better to inform people and encourage action because they help communities feel less vulnerable and more empowered. Collaborative partnerships with radio stations and other vehicles that reach into communities have a better chance of success. Partnerships developed as part of local projects enable scientists to interact with the public to share knowledge while establishing what issues resonate with a particular community. They can also address the community's anxiety and fears around climate change.

Scientists can therefore demonstrate that they are also human and are as concerned about the climate crisis as ordinary citizens. This helps to equip people with the information required to take action.

The Role of Environmental NGOs and How You Can Support Them

Many non-government organizations (NGOs) develop out of grassroots organizations, aiming to resolve situations that may affect them personally. Local leaders will champion the rest of the community and often engage with and put pressure on politicians. Attracting the attention of the media is a key activity, whether activists block roads for timber trucks, undertake marches, or picket outside law courts. Local groups generally aim to preserve the living conditions of communities that would be negatively impacted by the exploitation of natural resources.

Some NGOs collaborate with the government to ensure that environmental laws are enforced. They may also promote themselves as environmental experts to inform governments in this arena. NGOs sometimes assist with government initiatives so that they can ensure that positive environmental outcomes result. Lobbying by larger NGOs has been instrumental in addressing issues such as setting aside more land for conservation and informing legislators concerning the creation of policies and laws around issues such as air and water pollution, or the management of toxic waste or public lands. At the same time, NGOs need to strike a balance between collaboration and confrontation.

Corporate multinationals are sensitive to public perception and international NGOs can exploit this to enforce ethical environmental actions. NGOs are also often involved in certification programs and ethical ratings.

During the 1990s, NGOs started becoming involved in climate negotiations (Dalmedico & Buffet, 2000). Using analysis, journals, articles, and their relationships with their counterparts in developing countries, NGOs raised awareness of climate change risks among politicians and world leaders. NGOs also validated the credibility of climate science. NGOs frequently simplify the science so government officials and the general public understand the issues. NGOs work on the sidelines of COPs, submitting position papers and following working groups, while publishing daily bulletins to inform the public as to what is unfolding at the negotiation tables.

NGOs develop solutions, forming partnerships to formulate new proposals for negotiators, such as financial compensations for developing countries or forest management issues. Greenpeace, for example, presented a project aiming to stabilize emissions by 2050 (Dalmedico & Buffet, 2000).

One of the main ways that ordinary people can support NGOs in these efforts is through funding. In today's difficult economic environment, funding is crucial to enable NGOs to survive and advance the case for the environment in an age of climate change. Donations can take many forms besides finance—you could offer to provide skills free of charge or fund specific areas or activities.

Adapting to the New Normal

There's no doubt that the world has changed and the future isn't what it used to be. We are going to need to adopt new paradigms as our world changes, sometimes alarmingly quickly. However, there are positive steps that can be taken to help us adapt.

Designing Climate-Resilient Infrastructure

While infrastructure will often be negatively impacted by climate change impacts, it will also be an important part of building climate resilience. An extreme weather event such as a flood can have multiple impacts on infrastructure, besides affecting transport and electricity supplies in the affected region. Ensuring resilience reduces losses and the effects of disruption.

New infrastructure projects should account for climate change impacts, while existing infrastructure might need to be retrofitted. Additional infrastructure such as sea walls might need to be constructed. Traditional built infrastructure could be combined with natural barriers such as wetlands or floodplains. The public and private sectors will need to liaise to ensure climate resilience.

Building climate resilience could involve better management, as well as structural measures like raising bridge heights or enhancing natural drainage. Flexible, adaptive approaches could help to reduce the associated costs. Decisionmakers need to have access to high-quality information to plan effectively. Technical and institutional capacity should be increased to better manage climate-related risks.

Climate impacts are likely to spur infrastructure investment to mitigate the impacts of severe weather events, sea level rise, and others. Nature-based, flexible approaches may be more cost-effective than new builds or retrofitting infrastructure. The benefits of investing in resilient infrastructure far outweigh the associated costs. Developing infrastructure plans can also encourage investment. Climate resilience should be considered when considering competing bids for infrastructure projects.

As an example of retrofitting infrastructure, there are three approaches to address flooding and earthquake risks along the San Francisco waterfront. Defend, Accommodate, Retreat. In the case of the historic Ferry Building, and in the face of sea level rise, the Port of San Francisco has developed seven high-level Draft Waterfront Adaptation Strategies (SF Port, n.d.) through a collaborative five-year process. To defend San Francisco from three to seven feet of sea level rise expected by the year 2100, they are planning to raise the iconic, historic Ferry Building by as much as seven feet (SF Port, n.d.). Estimates to complete this project run in the billions of dollars.

Building Emotional Resilience in Communities

Climate change is unlikely to be resolved soon and holding out while the situation stabilizes means we need to become more emotionally resilient, so we're better able to cope with whatever the future brings. Eco-anxiety is becoming more common and can hinder efforts to take positive action. Below are a few strategies that can help to reduce feelings of apprehension or agitation:

- Check in with yourself regularly and recognize when you're starting to feel unsettled. Take a break from the mainstream news or focus on positive stories.
- Take time to identify and feel your emotions. It's normal to feel sad, angry, or anxious in our new climate change world. Be kind to yourself. None of us have ever been in this situation before and it's easy to feel overwhelmed.
- Press the pause button. Unplug for a while. Go for a walk, play with your children, or do something else you enjoy that makes you feel positive.
- Develop tools to help you cope when you start feeling anxious. Do deep-breathing exercises, pray or meditate, keep an emotional journal, or build gratitude into your day.
- Most people don't want to talk about global warming or even think about it. Find people you can talk to about your feelings—you'll likely find you're not the only one feeling this way.

- Several therapists and counselors now specialize in eco-anxiety and you might find it helpful to talk to a professional about how to cope with your feelings.
- Find hopeful stories. Read positive news and success stories. There's a lot being done worldwide to alleviate climate change impacts and there are some wonderful stories out there.
- Finally, if you're feeling overwhelmed, think about what you can do to improve things, including the ideas mentioned in this chapter. NGOs like Greenpeace and the World Wildlife Fund (WWF) also have climate action suggestions on their websites (Wessel, 2023).

Sustainable Choices Checklist

What can you do to save the planet? As mentioned in this chapter, there are a lot of positive actions you can take to slow climate change and reduce the emissions of your own household. The checklist below will provide you with some ideas of positive changes you could implement at home or the office.

- Drive less. Walk, bike, use public transport or carpool.
- Fly less and take non-stop flights when you do.
- Buy only what you need.
- Support eco-friendly brands.
- Buy secondhand clothing, furniture, and appliances.
- Repair rather than buying new.
- Donate items you no longer want or need to charities or worthy causes.
- Buy food produced locally or as close to your geographical area as possible.
- Buy organic food to support organic farmers.
- Implement Meat-Free Mondays and eat less meat.

- Compost your kitchen waste and garden clippings.
- Grow your own vegetables and fresh produce organically to avoid transport emissions.
- Buy in bulk and take your own containers for items like pasta, rice, nuts, candy, dried fruit, milk, fruit juice, and olive oil.
- Take reusable shopping bags to the store.
- Use less single-use plastic, including straws, plastic cups, cutlery, supermarket bags, etc. Plastic is produced using oil and natural gas, both of which are fossil fuels.
- Take a reusable coffee cup to your local cafe.
- Fill up a reusable water bottle at home instead of buying bottled water.
- Use reusable glasses, crockery, and flatware rather than disposables.
- Use cloth towels/rags in place of paper towels.
- Recycle paper, glass, tins, and plastic to save energy and reduce emissions. Be sure to wash your recyclables so they aren't contaminated.
- Only print out emails and documents when strictly necessary.
- Wear clothing made from natural materials like cotton, hemp, or bamboo. Polyester, viscose, and rayon are made using fossil fuels as a feedstock.
- Use your phone and laptop for as long as possible and only replace them when necessary to save on rare earth minerals and other components.
- Use energy-efficient lightbulbs like LEDs.
- Switch off lights when you leave a room.
- Unplug vampire electronics when not in use.
- Install insulation to maintain more even temperatures inside your home.
- Replace leaky windows with energy-efficient ones.

- Install solar panels or a hydropower system to provide your own electricity that emits no greenhouse gases.
- Switch to a tankless hot water heater.
- Use low-flow shower heads and toilets to save water.
- Wash full loads of laundry in the washing machine.
- Ensure the dishwasher is full before use.

Key Chapter Takeaways

- In this chapter, you will have discovered the things that you can do as an individual to make a difference. You can modify your home to use less energy and water, change your diet to reduce your carbon footprint and shop consciously rather than mindlessly buying random items.
- Reduce your home's waste stream by composting, recycling, repairing, and reusing wherever possible.
- Get involved in local conservation projects to help conserve wildlands or even parkland in your city or region. Alternatively, participate in a community garden project—or start one. Grow your own vegetables and fresh produce at home. Find out about grassroots movements in your area that are concerned about climate change and join their peaceful protests, awareness campaigns, or fundraising activities.
- Start a community dialogue to inform people about the realities of climate change and what they can do to reduce its impacts. Support environmental NGOs in their climate change awareness and mitigation efforts by donating funds or giving of your time and expertise.
- Adapting to our new normal means building climate-resilient infrastructure. Find ways to support local

initiatives and remember to climate-proof your own home. Finally, take care of yourself, especially if you suffer from eco-anxiety. Take a break from the news cycle, find friends or a therapist you can talk to, or take a break in the great outdoors.

- Imagine what a world united against climate change might look like—and set out an action plan as to how you could help make this happen.

Conclusion

Climate change affects everyone, regardless of social status, color, creed, or age. Extreme weather events, floods, droughts, wildfires, and sea level rise are impacting people across the globe, culminating in natural disasters that cause billions of dollars in damages in the US and elsewhere.

The Earth's atmosphere contains natural heat-trapping gases such as carbon dioxide, methane, nitrous oxide, and water vapor. When solar radiation reaches the Earth's surface, only a small portion escapes back into space. The rest is held in the atmosphere by these gases. This process keeps the planet warm enough to support life and is known as the greenhouse effect.

However, human beings have been releasing excessive greenhouse gases, from fossil fuel burning and industrial processes, into the atmosphere since the Industrial Revolution started around 200 years ago. Carbon dioxide is the most prevalent greenhouse gas but there are others as well, including gases released by novel chemicals used in refrigeration, semiconductor manufacturing, and other applications. This has caused an imbalance of these gases in the atmosphere, so temperatures on the Earth's surface are becoming hotter than normal. Although there are sometimes temperature anomalies that cause cooler seasons and years, the overall trend is upward. Both historic and current research indicate that fossil fuel burning is the primary driver of global warming.

Warming has now become significant enough to change the long-term weather patterns known as the climate. In the US, the Southeast is now primarily being affected by extreme weather such as tropical storms, hurricanes, and floods, while the Southwest has been enduring severe drought that sparks wildfires that destroy homes, livestock grazing, and business premises. Insurance companies are feeling the pinch and some have even gone bankrupt, with states needing to step in to honor damage claims after severe weather events. Insurance companies are reducing or excluding coverage for certain geographical locations, with state-run programs being obliged to fill the gap.

Sea level rise is bringing storm surges closer to land, resulting in devastating coastal floods, while low-lying atolls are gradually sinking into the ocean. As the ocean absorbs the excess carbon in the atmosphere, coral bleaches and shell-forming creatures can no longer create their homes.

Agriculture is being hard hit by changing weather patterns, particularly extreme heat, heavy precipitation, persistent drought, and floods. Growing seasons are changing and weather variability is affecting crop yields and the types of crops that can be grown. In some places, warmer weather means farmers can grow a greater variety of crops but it also makes for higher water usage. For livestock farmers, their animals cannot always cope with extreme temperatures; reproduction and productivity may be affected. There are concerns about a timing mismatch between blooming orchards and other crop plants and pollinators.

Commercial agriculture is a key greenhouse gas emitter, as farm equipment uses fossil fuels, nitrous oxide from inputs enters the atmosphere, and soil carbon is released due to unsustainable crop farming practices.

However, climate change affects more than human lives and activities. Across the globe, nature is on the move, attempting to escape rising temperatures, dwindling resources, disappearing habitats, wildfires, droughts, and floods. Changing ocean currents and temperatures have shifted fish and seafood species, affecting the fishing industry, tourism, and other ocean-based economic activities.

The 2023 summer was the hottest recorded in the Northern Hemisphere, with heat waves putting thousands of people in hospitals or sending them to doctors with heat-related ailments. Particulate pollution from fossil fuel burning is leading to respiratory diseases, heart complaints, type-2 diabetes, and dementia, particularly for people living near busy roads. Even children are affected, with babies more likely to develop respiratory complaints later in life.

People are traumatized, anxious, and stressed; some even struggle with PTSD after experiencing a climate-related disaster. Eco-anxiety has become a recognized psychological disorder that creates feelings of panic and distress.

But all is not lost. While climate change is here—and it is serious, threatening lives and livelihoods worldwide—there are many individuals, organizations, and initiatives working to stem the tide. At an international level, developed countries are being mandated to assist poor countries with funding to enable climate mitigation and adaptation. Signatories to the historic Paris Agreement have pledged to reduce emissions at scale, with many countries recently increasing their commitments as climate change bites. Organic farming is becoming more popular, as is the acceptance of renewable energy at all levels of society, with solar, wind, hydro, and geothermal resources becoming more prevalent as time goes on.

Technological solutions are also available to help us fight climate change. Carbon capture and storage (CCS) is already being done in the US and other countries. Another suggested method is to block or redirect sunlight by spraying sulfur-based aerosols into the atmosphere or seed clouds to reflect solar radiation back into space. These methods may have unintended consequences and could negatively affect the atmosphere, however, and might not be adopted.

Beacons of Hope

Precision agriculture is using technology to improve efficiency and yields while reducing inputs and fossil fuel usage. Farms using precision agriculture are experiencing increased yields and cost savings. Biotechnology is developing crops that use less water, are drought-tolerant, and have higher yields and nutritional value. Urban farming using hydroponic systems and vertical gardens is delivering fresh produce to city dwellers.

AI is being used to develop models that can forecast the weather and predict extreme weather events with greater accuracy. Trials are promising, although trained meteorologists are still required to verify results. AI can enable disaster preparation and management, as long as it is trained on the correct data sets.

In conservation, AI is tracking land-use change, species movements, and migrations, as well as animal behavior and how climate change is affecting wildlife, birds, and plants. AI is capable of rapidly analyzing reams of data, which means that trends can be assessed quickly and mitigation efforts put in place in a timely manner.

AI is improving electricity generation and reducing energy usage, particularly when employed to run smart grids, which supply electricity closer to where users are located. AI automatically moves electricity across the smart grid during outages, ensuring a steady supply of power for users. Several industrial applications are using AI to increase efficiency and reduce emissions. AI has reduced the quantity of alloys (new material) used in steel recycling, improved urban planning in an age of climate change, and reduced air conditioning requirements in buildings.

However, AI solutions are only as good as the data they are trained on and it is essential to ensure that the correct data sets are used to prevent unexpected or inaccurate outcomes, biases, and unintended privacy breaches. AI training needs to be appropriate for a particular situation to deliver the best results and accuracy.

Positive Case Studies

It is amazing what you can do when you put your mind to it. In 2006, the Ethiopian government introduced the Productive Safety Net Program to assist families who were unable to feed themselves due to environmental impacts. In exchange for taking part in local environmental programs such as land rehabilitation, improving water resources, or infrastructure construction, participants receive aid. Among other things, the program has reduced soil loss by over 40%, and increased land productivity and water availability (Pugacheva & Mrkaic, 2018).

In India, a decentralized air conditioning system where consumers are supplied with chilled water via underground pipes is reducing the high energy usage and pollution associated with regular air conditioners. It is already being rolled out in major cities. These centralized cooling systems, known as district cooling, consume 35%–50% less energy than individual air conditioning units (Pugacheva & Mrkaic, 2018).

Call to Action

You as an individual can do a great deal to slow climate change. Homes consume a significant amount of energy. This can be reduced by installing LED light bulbs, double-panel window glazing, ceiling fans, tankless water heaters, and even solar panels. Water conservation measures can be implemented too. Eating less meat, reducing food waste, composting, and supporting organic farmers will all help to save the planet. Becoming a conscious consumer, buying only what you need, will also help in the fight against climate change. You can also start a community food garden, join or support a grassroots movement in your area, help an environmental NGO with in-kind or financial donations, and talk to your community about climate change.

Go to a local Earth Day event on April 22, and learn about activities and programs in your area.

Please Leave a Review

We hope that you have found this book helpful in finding out more about climate change, as well as the tremendous international, regional, and local efforts being undertaken to slow the effects of global warming. If you have found this book helpful and inspirational, We would appreciate your leaving a review so others can benefit from this information.

We hope that the insights and strategies discussed have not only informed you but also inspired you to think differently about our planet and the role of technology in protecting it.

Here's your chance to make your voice heard and contribute to the global conversation on climate change and sustainability.

Your review can ignite Change and guide others in their journey towards environmental awareness and action. We're Counting on You to Spread the Word. You can inspire others to join the movement towards a healthier, more sustainable world.

By sharing your review, you're not just offering feedback; you're guiding others towards a path of awareness and action. Your voice can be the catalyst for someone else's environmental awakening.

Scan the QR code below to leave your review.

Thank you for joining us on this journey and for your commitment to our planet's future. Your insights are invaluable in helping others find the same help and hope you discovered.

References

Acciona. (n.d.). *Debunking climate change myths*. Www.activesustainability.com. https://www.activesustainability.com/climate-change/debunking-climate-change-myths

AI and the F6S community (2024 Feb 8) *65 Top Artificial Intelligence(AI) Companies and Startups in Austria in 2024* https://www.f6s.com/companies/artificial-intelligence/austria/co

Al Jazeera. (2023, March 30). *Rapidly melting Antarctic ice could affect oceans for centuries*. Www.aljazeera.com. https://www.aljazeera.com/news/2023/3/30/rapidly-melting-antarctic-ice-could-affect-oceans-for-centuries

Al Mubarak, R., & Alam, T. (2022, April 26). *The role of NGOs in tackling environmental Issues*. Middle East Institute. https://www.mei.edu/publications/role-ngos-tackling-environmental-issues

Andrews, R. G. (2023, August 10). *Is the Atlantic Ocean's great conveyer belt really slowing down?* Environment. https://www.nationalgeographic.com/environment/article/amoc-atlantic-ocean-conveyer-belt-climate-change

Apap, J. (2019). *The concept of climate refugee: Towards a possible definition*. European Parliamentary Research Service. https://www.europarl.europa.eu/RegData/etudes/BRIE/2018/621893/EPRS_BRI(2018)621893_EN.pdf

BBC World Service. (2023). *Where do we go when the seas rise?* BBC World Service. YouTube. https://www.youtube.com/watch?v=c0V1CtzFgrU

Begum, T. (2021, April 16). *Soil degradation: The problems and how to fix them*. Natural History Museum. https://www.nhm.ac.uk/discover/soil-degradation

Brown, D. (2022, February 7). *Is there a safe area of Florida with no chance of sinkholes?* Foundation Professionals of Florida. https://www.foundationprosfl.com/is-there-a-safe-area-of-florida-in-which-to-live-with-no-chance-of-sinkholes

Bruyninckx, H. (2023, August 29). *Clean water is life, health, food, leisure, energy…* European Environment Agency. https://www.eea.europa.eu/signals-archived/signals-2018-content-list/articles/clean-water-is-life-health

Buis, A. (2019, November 1). *The atmosphere: Keeping a weather eye on Earth's climate instabilities*. Climate Change: Vital Signs of the Planet. https://climate.nasa.gov/news/2918/the-atmosphere-keeping-a-weather-eye-on-earths-climate-instabilities

Buis, A. (2023, September 27). *Climate change evidence: How do we know?* Climate Change: Vital Signs of the Planet. https://climate.nasa.gov/evidence

Butler, R. (2023, September 8). *Deforestation in the Amazon rainforest continues to plunge.*

Mongabay Environmental News. https://news.mongabay.com/2023/09/deforestation-in-the-amazon-rainforest-continues-to-plunge

Cain, C. (2023, Nov 30) New Study Shows Liquefied Natural Gas Might be Worse for Climate Change than Coal. https://www.nbcbayarea.com/news/national-international/new-study-shows-liquefied-natural-gas-might-be-worse-for-climate-change-than-coal/3384705/#:~:text=By%20Chase%20-Cain%20%E2%80%A2%20Published,climate%20foot-print%20than%20burning%20coal.

Cain, D. (2023, July 9). *Wildlife conservation and AI*. Linkedin. https://www.linkedin.com/pulse/when-ai-met-bambi-tech-tale-wildlife-conservation-david-cain

CDP. (n.d.). *Cities at risk: Dealing with the pressures of climate change*. CDP. https://www.cdp.net/en/research/global-reports/cities-at-risk

Childs, J. W. (2023, November 1). *Maui wildfire death toll rises again*. The Weather Channel. https://weather.com/news/news/2023-11-01-maui-fire-death-toll-lahaina-hawaii

Cho, R. (2019, June 20). *How climate change impacts the economy*. Columbia Climate School. https://news.climate.columbia.edu/2019/06/20/climate-change-economy-impacts

Cianconi, P., Betro, S., & Janiri, L. (2020). *The impact of climate change on mental health: A systematic descriptive review*. Frontiers in Psychiatry, 11 – 2020. https://doi.org/10.3389/fpsyt.2020.00074

Client Earth. (2022, February 18). *Fossil fuels and climate change: the facts*. Client Earth. https://www.clientearth.org/latest/latest-updates/stories/fossil-fuels-and-climate-change-the-facts

Climate.gov. (2023, January 9). *Billion-dollar events to affect the United States from 1980 to 2022 (CPI-Adjusted)*. NOAA Climate.gov. https://www.climate.gov/media/14992

Community Conservation Inc. *What is community conservation?* (2021, October 1). Community Conservation Inc. https://communityconservation.org/our-work/what-is-community-conservation

Constellation. (n.d.). *35 Ways to reduce your carbon footprint*. Constellation. https://www.constellation.com/energy-101/energy-innovation/how-to-reduce-your-carbon-footprint

Cook, R. (2023, November 1). *Top 10 states with the most cattle*. Beef2Live. https://www.beef2live.com/story-top-10-states-cattle-0-110713

Cooley, A. (2021, April 22). *30 Day challenge to live an eco-friendly lifestyle*. Busch Systems. https://www.buschsystems.com/blog/news/30-day-challenge-live-an-eco-friendly-lifestyle

Dalmedico, A.D., & Buffet, C. (2000). *Environmental NGOs in the raising debate of climate change*. HAL SHS. https://shs.hal.science/halshs-00503072/document

Dangour, A. (2022, October 11). *How climate change affects vector-borne diseases*. Wellcome. https://wellcome.org/news/how-climate-change-affects-vector-borne-diseases

De Leon, S. P. (2019, March 20). *The role of smart grids and AI in the race to zero emissions*. Forbes. https://www.forbes.com/sites/cognitiveworld/2019/03/20/the-role-of-smart-grids-and-ai-in-the-race-to-zero-emissions

Delgado, G. (2022, July 27). *How can natural disasters produce social and cultural change?* Imana. https://www.imana.org/natural-disasters-produce-social-change

Desai, S. (2023, August 3). *The ultimate checklist for sustainable lifestyle*. Linkedin. https://www.linkedin.com/pulse/ultimate-checklist-sustainable-lifestyle-siddhesh-desai

DW Documentary. (2019). *Climate refugees in Bangladesh*. YouTube. https://www.youtube.com/watch?v=co5uywe-1Z8

ERS. (2023, October 31). *FAQs*. USDA Economic Research Service. https://www.ers.usda.gov/faqs

ESCAP. (n.d.). *Decentralized energy system fact sheet*. https://www.unescap.org/sites/default/files/14.%20FS-Decentralized-energy-system.pdf

Fletcher, C. (2023, July 24). *How to build climate-resilient communities*. Earth.org. https://earth.org/how-to-build-climate-resilient-communities

Frąckiewicz, M. (2023, August 28). *Harnessing AI for wildlife monitoring and habitat restoration*. TS2 SPACE. https://ts2.space/en/harnessing-ai-for-wildlife-monitoring-and-habitat-restoration

Fraisse, C., Ampatzidis, Y., Guzman, S., Lee, W., Martinez, C., Shukla, S., Singh, A., & Yu, Z. (2022, April 17). *AE571/AE571: Artificial intelligence (AI) for crop yield forecasting*. University of Florida IFAS Extension. https://edis.ifas.ufl.edu/publication/AE571

Freebody, P., & Tytler, R. (2023, June 12). *How should we teach climate change in schools? It starts with "turbo charging" teacher education*. The Conversation. https://theconversation.com/how-should-we-teach-climate-change-in-schools-it-starts-with-turbo-charging-teacher-education-207221

Ganyarpawar, T. (2023, March 9). *What are the main barriers to deploying large-scale energy storage systems?* Linkedin. https://www.linkedin.com/advice/1/what-main-barriers-deploying-large-scale-energy-storage

Global Partnership on AI. (2021). *Climate change and AI recommendations for government action*. GPAI. https://www.gpai.ai/projects/climate-change-and-ai.pdf

Glover, E. (2022, October 13). *8 Eco-friendly home improvements and how they make a difference*. Forbes Home. https://www.forbes.com/home-improvement/home/eco-friendly-home-improvements

Greenfield, N. (2022, May 9). *Climate migration and equity*. NRDC. https://www.nrdc.org/stories/climate-migration-equity

Hancock, L. (2023a). *Everything you need to know about coral bleaching—and how we can stop it*. World Wildlife Fund. https://www.worldwildlife.org/pages/everything-you-need-to-know-about-coral-bleaching-and-how-we-can-stop-it

Hancock, L. (2023b). *Why are glaciers and sea ice melting?* World Wildlife Fund. https://www.worldwildlife.org/pages/why-are-glaciers-and-sea-ice-melting

Harvey, F. (2021, January 25). *Global ice loss accelerating at record rate, study finds.* The Guardian. https://www.theguardian.com/environment/2021/jan/25/global-ice-loss-accelerating-at-record-rate-study-finds

Hays, K. (2023, October 10). *The world's first real AI rules are coming soon. Here's what they will look like.* Business Insider. https://www.businessinsider.com/ai-regulation-2023-us-eu-china-100-10

Hill, A. C. (2023, August 17). *Climate change and U.S. property insurance: A stormy mix.* Council on Foreign Relations. https://www.cfr.org/article/climate-change-and-us-property-insurance-stormy-mix

Howarth, R.W.(2023, Oct 24) Cornell University *The Greenhouse gas footprint of Liquefied Natural Gas (LNG) Exported from the United States https://www.research.howarthlab.org/publications/Howarth_LNG_assessment_preprint_archived_2023-1103.pdf*

Huang, H. (2022, October 27). *Fossil fuels lobby.* Encyclopedia.pub. https://encyclopedia.pub/entry/31512

Iberdrola. (2020, January 21). *Green jobs: Good for you, the environment, and for the economy.* Iberdrola. https://www.iberdrola.com/sustainability/what-are-green-jobs

International Federation of the Red Cross. (2020). *Tackling the humanitarian impacts of the climate crisis together: World disasters Report 2020.* IFRC.org. https://www.ifrc.org/sites/default/files/2021-09/IFRC_WDR_ExecutiveSummary_EN_Web.pdf

International Federation of the Red Cross. (2022). *Where it matters most: Smart climate financing for the hardest hit people.* IFRC.org. https://www.ifrc.org/sites/default/files/2022-11/20221108_ClimateSmartFinance.pdf

ITU News. (2022, March 24). *How AI can help protect us from disasters.* ITU Hub. https://www.itu.int/hub/2022/03/ai-disaster-management-early-warning

IUCN. (2023, July 20). *Computer conservation.* IUCN. https://www.iucn.org/story/202307/computer-conservation

Jones, E., & Easterday, B. (2022, June 28). *Artificial Intelligence's Environmental Costs and Promise.* Council on Foreign Relations. https://www.cfr.org/blog/artificial-intelligences-environmental-costs-and-promise

Kelly, F. J., & Fussell, J. C. (2015). *Air pollution and public health: Emerging hazards and improved understanding of risk.* Environmental Geochemistry and Health, *37*(4), 631–649. https://doi.org/10.1007/s10653-015-9720-1

Kelly, R., Nettlefold, J., Mossop, D., Bettiol, S., Corney, S., Cullen-Knox, C., Fleming, A., Leith, P., Melbourne-Thomas, J., Ogier, E., van Putten, I., & Pecl, G. T. (2020). *Let's talk about climate change: Developing effective conversations between scientists and communities.* One Earth, *3*(4), 415–419. https://doi.org/10.1016/j.oneear.2020.09.009

Klein, N. (2015 August 4) This Changes Everything: Capitalism vs The Climate https://thischangeseverything.org/book/

Kreier, F. (2021, November 5). *Earth's lower atmosphere is rising due to climate change.* Science News. https://www.sciencenews.org/article/earth-lower-atmosphere-rising-climate-change-troposphere

Kuta, S. (2023, February 10). *Fifteen million people at risk of severe floods from melting glaciers.* Smithsonian Magazine. https://www.smithsonianmag.com/smart-news/fifteen-million-people-at-risk-of-severe-floods-from-melting-glaciers-180981608

Land Stewardship Center. (n.d.). *The green communities guide: Advancing nature-based solutions.* Green Communities Guide. https://greencommunitiesguide.ca/guide

Lemmons, R. (2023, March 20). *The natural and enhanced greenhouse effect—Greenhouse gases.* Climate Policy Watcher. https://www.climate-policy-watcher.org/greenhouse-gases-2/the-natural-and-enhanced-greenhouse-effect.html

Lin, Z. (2021, August 9). *Record-breaking extreme heat and cold events may result from similar underlying mechanisms.* EurekAlert. https://www.eurekalert.org/news-releases/924543

Lindsey, R. (2023, May 12). *Climate change: Atmospheric carbon dioxide.* NOAA Climate.gov. https://www.climate.gov/news-features/understanding-climate/climate-change-atmospheric-carbon-dioxide

Longley, R. (2020, October 30). *What is a grassroots movement? Definition and examples.* ThoughtCo. https://www.thoughtco.com/grassroots-movement-definition-and-examples-5085222

Loubere, P. (2012). *The global climate system.* The Nature Education Knowledge Project. https://www.nature.com/scitable/knowledge/library/the-global-climate-system-74649049/

Mammoth Discovery. *When were the ices ages and why are they called that?* (2019). Mammoth Discovery. https://www.cdm.org/mammothdiscovery/wheniceages.html

Mandy. (2023, March 17). *What is coral bleaching and why should you care?* Coral Reef Alliance. https://coral.org/en/blog/what-is-coral-bleaching-and-why-should-you-care

Marchant, N. (2021, June 28). *This is how climate change could impact the global economy.* World Economic Forum. https://www.weforum.org/agenda/2021/06/impact-climate-change-global-gdp

Maycock, T. (2022, September). *Billion-dollar disasters are happening more often.* North Carolina Institute for Climate Studies. https://ncics.org/cics-news/billion-dollar-disasters-are-happening-more-often

McGovern, A., Ebert-Uphoff, I., Gagne, D. J., & Bostrom, A. (2022, April 13). *Why we need to focus on developing ethical, responsible, and trustworthy artificial intelligence approaches for environmental science.* Environmental Data Science, 1. https://doi.org/10.1017/eds.2022.5

McNally, A. (2020, May 26). *Creating trustworthy AI for the environment: Transparency, bias, and beneficial use*. Markkula Center for Applied Ethics | Santa Clara University. https://www.scu.edu/environmental-ethics/resources/creating-trustworthy-ai-for-the-environment-transparency-bias-and-beneficial-use

Mélanie. (2021, February 1). *Mangroves: natural barrier against rising waters*. CLS. https://www.cls.fr/en/mangroves-natural-barrier-against-rising-waters

Mercy Corps. (2021, April 8). *The facts: How climate change affects people living in poverty*. Mercy Corps. https://www.mercycorps.org/blog/climate-change-poverty

Meyer, K., & Newman, P. (2020). *The Holocene, the Anthropocene, and the Planetary Boundaries*. Planetary Accounting, 35–52. https://doi.org/10.1007/978-981-15-1443-2_3

Miller, R.G. (2006) Inside Global Warming *Science Scope, 30 (2), 56-60 http://digital commons.chapman.edu/education_articles/13?utm_source=digitalcommons.chapman.edu%2Feducation_articles%2F13&utm_medium=PDF&utm_campaign=PDFCoverPages*

Montague, M. (2021, July 2). *Could mangroves help save the world?* BBC Earth. https://www.bbcearth.com/news/could-mangroves-help-save-our-planet

Mooney, C. (2023, January 5). *Half of Earth's glaciers could melt even if key warming goal is met, study says*. Washington Post. https://www.washingtonpost.com/climate-environment/2023/01/05/glaciers-melt-this-century-warming

Moore, H. (2020, March 27). *Why grassroots activism is required to change climate policy*. NB Media Co-Op. https://nbmediacoop.org/2020/03/27/why-grassroots-activism-is-required-to-change-climate-policy

Mulhern, O. (2020, November 20). *How satellites help tackle climate change*. Earth.org. https://earth.org/data_visualization/how-satellites-help-tackle-climate-change

National Academy of Sciences. (2021, August 12). *Global warming is contributing to severe weather events*. https://www.nationalacademies.org/based-on-science/climate-change-global-warming-is-contributing-to-extreme-weather-events

National Grid. (2023, February 28). *What is carbon capture and storage?* https://www.nationalgrid.com/stories/energy-explained/what-is-ccs-how-does-it-work

National Weather Service. (2019). *Flash Flooding Definition*. Weather.gov. https://www.weather.gov/phi/FlashFloodingDefinition

NOAA. (2009, October 2). *What is the difference between weather and climate?* National Oceanic and Atmospheric Administration. https://oceanservice.noaa.gov/facts/weather_climate.html

NOAA. (2011, February 1). *Weather systems and patterns*. National Oceanic and Atmospheric Administration. https://www.noaa.gov/education/resource-collections/weather-atmosphere/weather-systems-patterns

NOAA. (2021, August 13). *Climate change impacts*. National Oceanic and Atmospheric Administration. https://www.noaa.gov/education/resource-collections/climate/climate-change-impacts

NOAA. (2023a, January 20). *What is Ocean Acidification?* National Oceanic and

Atmospheric Administration. https://oceanservice.noaa.gov/facts/acidification.html

NOAA. (2023b, October 6). *Heavy rains cause flooding in New York City*. National Oceanic and Atmospheric Administration. https://www.ncsdis.noaa.gov/news/heavy-rains-cause-flooding-new-york-city

Nova PBS Official. (2022, February 3). *Arctic sinkholes*. YouTube. https://www.youtube.com/watch?v=HvKpnaXYUPU

Nowacki, I. (2023, October 30). *How is AI assisting in optimizing renewable energy sources?* TS2 Space. https://ts2.space/en/how-is-ai-assisting-in-optimizing-renewable-energy-sources/

OECD. (2018). *Climate-resilient infrastructure*. OECD. https://www.oecd.org/environment/cc/policy-perspectives-climate-resilient-infrastructure.pdf

Oladipo, G. (2023, September 11). *US sets new record for billion-dollar climate disasters in single year*. The Guardian. https://www.theguardian.com/environment/2023/sep/11/us-record-billion-dollar-climate-disasters

Pelling, M. (2007). *Urbanization and disaster risk panel contribution to the population-environment research network cyberseminar on population and natural hazards*. https://citeseerx.ist.psu.edu/document?repid=rep1&type=pdf&doi=c64c27e09397149d4fdb7997ace325571275782e

Phillips, A. (2023, November 24). *Iceberg five times the size of New York breaks away from Antarctica*. Newsweek.com https://www.newsweek.com/iceberg-five-times-size-new-york-breaks-away-antarctica-1846793

Pokhrel, A. (2023, July 19). *How can you incorporate sustainability into disaster recovery?* Linkedin. https://www.linkedin.com/advice/0/how-can-you-incorporate-sustainability-disaster

Pontifical Academy of Sciences. (n.d.). *Science education and climate change*. Pontifical Academy of Sciences. https://www.pas.va/en/publications/acta/acta25pas/quere.html

Prange, M. (2022, December 19). *Climate change is fueling migration. Do climate migrants have legal protections?* Council on Foreign Relations. https://www.cfr.org/in-brief/climate-change-fueling-migration-do-climate-migrants-have-legal-protections

Pugacheva, E., & Mrkaic, M. (2018, March 20). *Adapting to climate change—three success stories*. IMF. https://www.imf.org/en/Blogs/Articles/2018/03/20/adapting-to-climate-change-three-success-stories

Ratcliff, A. (2020, September 23). *Carbon emissions of richest 1 percent more than double the emissions of the poorest half of humanity*. Oxfam International. https://www.oxfam.org/en/press-releases/carbon-emissions-richest-1-percent-more-double-emissions-poorest-half-humanity

Resilient Cities Network. (2018, October 7). *What is urban resilience?* Resilient Cities Network. https://resilientcitiesnetwork.org/what-is-urban-resilience

Reuters. (2021, June 9). *Keystone pipeline officially canceled after Biden revokes key permit.*

CNBC. https://www.cnbc.com/2021/06/09/tc-energy-terminates-keystone-xl-pipeline-project.html

Ritchie, H. (2020, October 6). *Cars, planes, trains: Where do CO2 emissions from transport come from?* Our World in Data. https://ourworldindata.org/co2-emissions-from-transport

Ross, H. (2021, August 7). *Climate risks for insurers: Why the industry needs to act now to address climate risk on both sides of the balance sheet.* S&P Global. https://www.spglobal.com/esg/insights/climate-risks-for-insurers-why-the-industry-needs-to-act-now-to-address-climate-risk-on-both-sides-of-the-balance-sheet

Rubin, E. S. (2018, June 21). *Innovation and climate change.* BBVA Open Mind. https://www.bbvaopenmind.com/en/articles/innovation-and-climate-change

Ruivivar, L. (2021, July 22). *5 Uses of AI in emergency management.* Havrion. https://havrion.com/5-uses-of-ai-in-emergency-management

Russell, C. (2022, December 2). *Climate change in the media: Public perception and the responsibility of news outlets.* Earth Day. https://www.earthday.org/climate-change-in-the-media-public-perception-and-the-responsibility-of-news-outlets

Santander Open Academy. (2023, September 8). *Why is conscious consumerism the key to a sustainable future?* Santander Open Academy. https://www.becas-santander.com/en/blog/conscious-consumerism.html

Schewel, K. (2023, July 20). *Who counts as a climate migrant?* ReliefWeb. https://reliefweb.int/report/world/who-counts-climate-migrant

Scott, D. (2012). *Geoengineering and environmental ethics.* The Nature Education Knowledge Project. https://www.nature.com/scitable/knowledge/library/geoengineering-and-environmental-ethics-80061230

Seervai, S., Gustafsson, L., & Abrams, M. K. (2022, May 4). *The impact of climate change on our health and health systems.* Commonwealth Fund. https://www.commonwealthfund.org/publications/explainer/2022/may/impact-climate-change-our-health-and-health-systems

Sejian, V., Gaughan, J. B., Bhatta, R., & Naqvi, S. M. K. (2016). *Impact of climate change on livestock productivity.* Journal of Animal Science, 94 (suppl_5), 619–620. https://doi.org/10.2527/jam2016-1284

Seman, S. (n.d.). *Natural drivers of climate change.* Meteo 3: Introductory Meteorology. https://www.e-education.psu.edu/meteo3/node/2246

Serin, E. (2022, June 7). *Can we have economic growth and tackle climate change at the same time?* Grantham Research Institute on Climate Change and the Environment. https://www.lse.ac.uk/granthaminstitute/explainers/can-we-have-economic-growth-and-tackle-climate-change-at-the-same-time

SF Port. (n.d.a). *Defending San Francisco's waterfront from flood risks.* SF Port. https://sfport.com/wrp/adaptation-strategies

SF Port. (n.d.b). *Draft waterfront adaptation strategies.* SF Port. https://sfport.com/wrp/waterfront-adaptation

Shade, J., & Tully, K. (n.d.). *Organic farming practices for improving soil health.* Organic

Center. https://www.organic-center.org/sites/default/files/project/2020/03/soil-health-review_shadetully.pdf

Shaw, R. (2023, September 25). *A climate high, a climate low, and our climate future.* World Wildlife Fund. https://www.worldwildlife.org/stories/a-climate-high-a-climate-low-and-our-climate-future

Sherriff, L. (2023, September 5). *Babcock Ranch: Florida's first hurricane-proof town.* BBC News. https://www.bbc.com/future/article/20230904-babcock-ranch-floridas-first-hurricane-proof-town

Skarbek, A. (2022, May 6). *Tackling climate change requires university, government and industry collaboration—here's how.* The Campus. https://www.timeshighereducation.com/campus/tackling-climate-change-requires-university-government-and-industry-collaboration-heres-how

Smith, A. B. (2023, January 10). *2022 U.S. billion-dollar weather and climate disasters in historical context.* NOAA Climate.gov. https://www.climate.gov/news-features/blogs/beyond-data/2022-us-billion-dollar-weather-and-climate-disasters-historical

Solis-Moreira, J. (2023, July 7). *AI forecasts could help us plan for a world with more extreme weather.* Popular Science. https://www.popsci.com/environment/ai-weather-prediction-accuracy

Sparks & Honey. (n.d.). *Gen Z and climate.* Sparks & Honey. https://www.sparksandhoney.com/gen-z-complexities-climate

Statista. (2023, June). *Top 10 US states by inventory of hogs and pigs as of March 2023 (in 1,000s).* https://www.statista.com/statistics/194371/top-10-us-states-by-number-of-hogs-and-pigs

Szuwalski, Aydin, Fedena, Garber-Yonts, Litzow (2023, October 19) *The collapse of eastern Bering Sea snow crab.* Science, Vol 382 Issue 6668 pp 306–310 Science.org. https://www.science.org/doi/10.1126/science.adf6035

Task Force on Climate-Related Financial Disclosure (2023 Nov) https://www.fsb-tcfd.org/

Texas Monthly. (2018, August 24). *Voices from the storm, a year later.* Texas Monthly. https://www.texasmonthly.com/news-politics/harvey-survivors-one-year-later/

The Royal Society. (2013). *The basics of climate change.* The Royal Society. https://royalsociety.org/topics-policy/projects/climate-change-evidence-causes/basics-of-climate-change

The Royal Society. (2020, March). *How does climate change affect the strength and frequency of floods, droughts, hurricanes, and tornadoes?* The Royal Society https://royalsociety.org/topics-policy/projects/climate-change-evidence-causes/question-13

The Royal Society. (2022). *How does deforestation affect biodiversity?* The Royal Society. https://royalsociety.org/topics-policy/projects/biodiversity/deforestation-and-biodiversity/

The White House. (2022, September 1). *The rising costs of extreme weather events*. The White House. https://www.whitehouse.gov/cea/written-materials/2022/09/01/the-rising-costs-of-extreme-weather-events

The Wilderness Society. (2022, October 21). *5 stories of people impacted by climate change and inspired to take action*. https://www.wilderness.org/articles/blog/5-stories-people-impacted-climate-change-and-inspired-take-action

Tiernan, J. (2019, February 11). *Future farms: Agritech innovations to feed a changing planet*. The Beam Magazine. https://medium.com/thebeammagazine/future-farms-agritech-innovations-to-feed-a-changing-planet-b6e69666a237

Tree Plantation. (2023, March 3). *The lungs of the earth may soon be out of breath*. Tree Plantation. https://treeplantation.com/amazon-rainforest.html

UCAR Center for Science Education. (n.d.). *Climate variability*. UCAR Center for Science Education. https://scied.ucar.edu/learning-zone/how-climate-works/climate-variability

UK Research and Innovation. (2023). *A brief history of climate change discoveries*. https://www.discover.ukri.org/a-brief-history-of-climate-change-discoveries

United Nations. (n.d.). *All About the NDCs*. https://www.un.org/en/climatechange/all-about-ndcs

United Nations Climate Change. (2023). *Introduction to climate finance*. Unfccc.int. https://unfccc.int/topics/introduction-to-climate-finance

United Nations Development Program. (2021, January 17). *7 things we have learned about adapting to climate change*. UNDP. https://undp-climate.exposure.co/7-lessons-on-adaptation

United Nations Development Program. (2023, June 30). *Climate change is a matter of justice—here's why*. UNDP Climate Promise. https://climatepromise.undp.org/news-and-stories/climate-change-matter-justice-heres-why

United Nations Development for Risk Reduction. (2021, June 9). *Poverty and inequality*. UNDRR PreventionWeb. https://www.preventionweb.net/understanding-disaster-risk/risk-drivers/poverty-inequality

UN Food and Agricultural Organization. (2016, April 15). *Disaster risk reduction (DRR) and climate change adaptation (CCA)*. FAO. https://www.slideshare.net/FAOoftheUN/disaster-risk-reduction-drr-and-climate-change-adaptation-cca

UNFCCC. (2023). *Synthesis report on existing funding arrangements and innovative sources relevant to addressing loss and damage associated with the adverse effects of climate change*. Transitional Committee Second meeting of the Transitional Committee. https://unfccc.int/sites/default/files/resource/TC2_SynthesisReport23May23.pdf

UNHCR. (2022). *Climate change and disaster displacement*. UNHCR. https://www.unhcr.org/what-we-do/build-better-futures/environment-disasters-and-climate-change/climate-change-and

USDA Climate Hubs. (2013). *Growing seasons in a changing climate*. USDA. https://www.climatehubs.usda.gov/growing-seasons-changing-climate

US Department of the Interior. (2021, May 6). *America the Beautiful: Our work to conserve at least 30% of lands and waters by 2030*. US Department of the Interior. https://www.doi.gov/blog/america-beautiful-our-work-conserve-least-30-lands-and-waters-2030

US Environmental Protection Agency. (n.d.). *Climate impacts on agriculture and food supply*. US EPA. https://climatechange.chicago.gov/climate-impacts/climate-impacts-agriculture-and-food-supply

US Environmental Protection Agency. (2022, December 11). *Climate change impacts on the ocean and marine resources*. US EPA. https://www.epa.gov/climateimpacts/climate-change-impacts-ocean-and-marine-resources

USA Facts. (2023, March 28). *Are major natural disasters on the rise due to climate change?* USA Facts. https://usafacts.org/articles/are-the-number-of-major-natural-disasters-increasing

US Global Change Research Program. (2010). *Climate Science Special Report* 1, 1–470. https://science2017.globalchange.gov/chapter/1/

Vaia. (n.d.). *The atmosphere: Importance & composition*. Vaia. https://www.hellovaia.com/explanations/environmental-science/physical-environment/the-atmosphere

Van Rijmenam, M. (2023, February 16). *Privacy in the age of AI: Risks, challenges and solutions*. The Digital Speaker. https://www.thedigitalspeaker.com/privacy-age-ai-risks-challenges-solutions

Voiland, A. (2023, January 4). *Weather whiplash*. NASA Earth Observatory. https://earthobservatory.nasa.gov/images/150797/weather-whiplash

Volunteer FDIP. (2023, August 16). *Top 10 environmental celebrity activists inspiring the world to volunteer with climate action to stop climate change*. https://www.volunteerfdip.org/environmental-celebrity-activists-inspiring-the-world-to-volunteer-with-climate-action-to-stop-climate-change

Water Science School. (2018, June 6). *Groundwater decline and depletion*. US Geological Survey. https://www.usgs.gov/special-topics/water-science-school/science/groundwater-decline-and-depletion

Webb, L. (2023, June 12). *Green AI: The role of artificial intelligence in energy conservation*. Caribbean Electric Utility Services Corporation. https://www.carilec.org/green-ai-the-role-of-artificial-intelligence-in-energy-conservation

Wessel, M. (2023, August 10). *What is eco-anxiety?* Charity Digital. https://charitydigital.org.uk/topics/topics/climate-change-how-to-build-resilience-11135

Westcott, M. (2021). *Tourism and climate change*. Pressbooks. https://opentextbc.ca/introtourism2e/chapter/tourism-and-climate-change

Wikipedia. (2013, January 11). *Global warming. Short-term variations versus a long-term trend*. Wikipedia. Commons. https://en.m.wikipedia.org/wiki/File:Global_warming._Short-term_variations_versus_a_long-term_trend_(NCADAC).png

Wikipedia Contributors. (2023a, July 7). *Thermohaline circulation*. Wikimedia Foundation. https://en.wikipedia.org/wiki/Thermohaline_circulation

Wikipedia Contributors. (2023b, October 4). *Effects of climate change on small island countries*. Wikipedia. https://en.wikipedia.org/wiki/Effects_of_climate_change_on_small_island_countries

Wikipedia Contributors. (2023c, October 23). *Camp Fire (2018)*. Wikipedia Foundation. https://en.wikipedia.org/wiki/Camp_Fire_(2018)

Wikipedia Contributors. (2023d, November 1). *Extinction Rebellion*. Wikimedia Foundation. https://en.wikipedia.org/wiki/Extinction_Rebellion

Wikipedia Contributors. (2023e, November 7). *Hurricane Katrina*. Wikimedia Foundation. https://en.wikipedia.org/wiki/Hurricane_Katrina

Williams, G. (2023, June 22). *Solar, wind, or hydro energy—which renewable is right for you?* Purelight Power. https://purelightpower.com/blog/national/solar-wind-hydro-renewable-energy

World Population Review. (2023). *Chicken production by state, 2023*. https://worldpopulationreview.com/state-rankings/chicken-production-by-state

World Weather Attribution. (2023, July 25). *Extreme heat in North America, Europe and China in July 2023 made much more likely by climate change*. https://www.worldweatherattribution.org/extreme-heat-in-north-america-europe-and-china-in-july-2023-made-much-more-likely-by-climate-change

Ye, J. (2023, February 3). *Main greenhouse gases*. Center for Climate and Energy Solutions. https://www.c2es.org/content/main-greenhouse-gases

Made in the USA
Las Vegas, NV
13 April 2024

88626390R00111